MW00355406

FUZZY SET THEORY
Foundations and Applications

George J. Klir
Ute St.Clair
Bo Yuan

To join a Prentice Hall PTR Internet
mailing list, point to:

http://www.prenhall.com/mail_lists/

Prentice Hall PTR
Upper Saddle River, NJ 07458
http://www.prenhall.com

Library of Congress Cataloging-in-Publication Data
Klir, George J.
    Fuzzy set theory : foundations and applications / George J. Klir,
Ute St. Clair, Bo Yuan.
        p.   cm.
    Includes bibliographical references and index.
    ISBN: 0-13-341058-7 (alk. paper)
    1. Fuzzy sets.   I. St. Clair, Ute H.   II. Yuan, Bo.   III. Title.
QA248.5.K57   1997
511.3'22--dc21   96-40032
                                                           CIP

Editorial/production supervision and design: *Patti Guerrieri*
Cover design director: *Jerry Votta*
Cover designer: *Ute St. Clair*
Manufacturing manager: *Alexis R. Heydt*
Marketing manager: *Dan Rush*
Acquisitions editor: *Paul W. Becker*
Editorial assistant: *Maureen Diana*

 ©1997 by Prentice Hall PTR
Prentice-Hall, Inc.
A Simon & Schuster Company
Upper Saddle River, NJ 07458

The publisher offers discounts on this book when ordered in bulk quantities. For more information, contact: Corporate Sales Department, Prentice Hall PTR, One Lake Street, Upper Saddle River, NJ 07458, Phone: 800-382-3419, Fax: 201-236-7141, e-mail: corp-sales@prenhall.com

All products or services mentioned in this book are the trademarks or service marks of their respective companies or organizations.

Printed in the United States of America
10 9 8 7 6 5 4 3 2 1

ISBN   0-13-341058-7

Prentice-Hall International (UK) Limited, *London*
Prentice-Hall of Australia Pty. Limited, *Sydney*
Prentice-Hall Canada Inc., *Toronto*
Prentice-Hall Hispanoamericana, S.A., *Mexico*
Prentice-Hall of India Private Limited, *New Delhi*
Prentice-Hall of Japan, Inc., *Tokyo*
Simon & Schuster Asia Pte. Ltd., *Singapore*
Editora Prentice-Hall do Brasil, Ltda., *Rio de Janeiro*

CONTENTS

PREFACE

Since the early 1990s, literature on fuzzy set theory and its various applications has been rapidly growing. Hundreds of books on this subject are now available on the market. Most of them are either edited collections of papers on various themes or monographs on special topics. Textbooks on fuzzy set theory are still rather rare, in spite of the growing need for such textbooks at all levels of higher education.

This book, *Fuzzy Set Theory: Foundations and Applications*, is intended to fill a particular gap in the literature. Its aim is to serve as a textbook for a general course in undergraduate liberal arts and sciences programs. This aim is reflected in the content of the book and the style in which it is written.

As the title of the book suggests, it is a simple introduction to basic elements of fuzzy set theory. However, it also contains an overview of the corresponding elements of classical set theory, including basic ideas of classical relations, as well as an overview of classical logic. The emphasis is on conceptual rather than theoretical presentation of the material.

We hope the text will help to develop undergraduate courses in which students will be exposed at an early stage of their studies to basic ideas of the increasingly important fuzzy set theory. The course may already be offered at the freshman or sophomore levels. It may be viewed as a meaningful replacement and enrichment of a typical course on the basics of classical set theory and classical logic, which is often required in undergraduate liberal arts and sciences programs.

Since the text is intended for a general course and not a course oriented to majors in any particular discipline, it covers only the most fundamental concepts of fuzzy set theory. Various special topics, particularly relevant to some disciplines, are not covered. Examples are chosen, by and large, from daily life. The broad applicability of fuzzy set theory in many areas of human affairs is surveyed but not covered in detail. Only a few examples of applications are covered, chosen again from daily life.

For further study of fuzzy set theory and fuzzy logic, the graduate text *Fuzzy Sets and Fuzzy Logic: Theory and Applications* by G. J. Klir and B. Yuan (Prentice Hall, 1995) offers the most natural continuation of this undergraduate text. The main advantage is that both books use the same terminology and notation.

We expect that any shortcomings of this text, caused either by its content or its style, will emerge from its use in various classroom environments. We intend to identify them, hopefully with the help of instructors and students using the text, and make appropriate revisions at some time in the future. For this purpose, we welcome any suggestions from instructors, students, or other readers of the book.

Lastly, we would like to express our gratitude to Ms. Deborah Stungis for reading the entire manuscript of this book and for providing us with many useful comments, as well as to Ms. Yin Pan for composing parts of the book and contributing to various technical issues.

George J. Klir, Ute St.Clair, and Bo Yuan
Binghamton, New York

1

INTRODUCTION

1.1 INFORMATION, UNCERTAINTY, AND COMPLEXITY

We human beings are intelligent agents trying to make sense of our environment and to make plans to direct our lives in accordance with our needs and wishes. To achieve these ends, we make use of the knowledge we gain from experiencing the world within which we live and use our ability to reason to create order in the mass of information at our disposal. In ordinary life, we use this information to understand our surroundings, to learn new things, and to make plans for the future. Thus, we have developed the ability to reason on the basis of evidence in order to achieve our goals. Of course, since we are all limited in our ability to perceive the world and limited in how profoundly we reason, we find ourselves everywhere confronted by uncertainty: uncertainty about the adequacy of our information and uncertainty about how good our inferences are. *Uncertainty* is the condition in which the possibility of error exists, because we have less than total information about our environment.

To survive in our world, we are engaged in making decisions, managing and analyzing information, as well as predicting future events. All of these activities utilize information that is available and help us try to cope with information that is not. Lack of information, of course, produces uncertainty. But this interplay of information and

1

uncertainty is the hallmark of complexity. One example illustrating a common experience with complexity is this description of driving a car.

We can probably agree that driving a car is (at least relatively) complex. Further, driving a standard transmission or stick-shift car is more complex than driving a car with an automatic transmission, one index of this being that more description is needed to cover adequately our knowledge of driving in the former case than in the latter (we must know, for instance, the revolutions per minute of the engine and how to use the clutch). Thus, because more knowledge is involved in the driving of a standard-transmission car, it is more complex. However, the complexity of driving also involves the degree of our uncertainty; for example, we do not know precisely when we will have to stop or swerve to avoid an obstacle. As our uncertainty increases—for instance, in heavy traffic or on unfamiliar roads—so does the complexity of the task. Thus our perception of complexity increases both when we realize how much we know and when we realize how much we do not know.

The most important problem of coming to know is therefore the problem of how to make complexity less daunting. This requires that we make appropriate simplifications, which seem to be best achieved in each particular situation by making a satisfactory compromise between the information we have and the amount of uncertainty we are willing to accept.

1.2 MEASUREMENT AND UNCERTAINTY

We are all familiar with empirical investigations, such as scientific explorations and criminal investigations, in which scientists must examine physical events and objects in order to solve a problem or to gain knowledge about some phenomenon. For example, in many criminal cases, forensic scientists examine minute traces of blood in order to discover the unique DNA identification of the sample. This kind of investigation requires a great deal of precision, partly because the samples are often quite small, and partly because small deviations in the accuracy of measurement of microscopic elements, such as DNA sequences, can result in misidentification. Thus, scientists dealing with physical substances, in particular, must be concerned with maintaining a high standard of precision.

However, even when we acknowledge this, we must also be alert to the limits of the quest for precision. First, no matter how accurate our measurements are, some uncertainty always remains, even if it is of a very small magnitude. All our measurements are taken with reference to some artificial standard; for example, we have accepted a standard for length—the foot—which is further divided into smaller units, namely inches. These, in turn, are divided on our measuring sticks into fractional parts. We can say that these fractional parts are the smallest units we can distinguish on our measuring instruments. Measuring sticks used for practical applications usually distinguish successive units of one-sixteenth inch each. For a laboratory scientist, however, this distinction might be too coarse, and so he or she might use an instrument that distinguishes units of one sixty-fourth of an inch. In the case of metric measurements, we know that sophisticated instruments can measure down to one-millionth of a meter, because the objects studied are microscopically small. Further, when we concentrate on the smallest interval distinguished by our measuring instrument, we see that the measurement taken of a tiny object might fall between two ends of even such a small unit; in other words, we can divide the one-millionth unit even more—indeed, there is no theoretical limit to the number of times we can divide a measurement interval. However, in practice, we are able to make only a finite number of divisions. This means ultimately that all our measurements, no matter how precise, admit the possibility of error: Even the most precise measurements are uncertain!

Thus, there exists a residue of uncertainty in the hardest sciences, those in which the data investigated are most amenable to precise description and measurement. If this is true in these sciences, then how much more does our insight apply to sciences whose objects of study cannot be so readily quantified? In the social sciences, for example, the objects of study are either individual human beings or groups of them. They are more complex than any inanimate objects, or even any other animals. Scientists in the fields of psychology and sociology have attempted to impose scientific rigor onto their objects of study and have often succeeded to an admirable degree. However, the low uncertainty enjoyed by physicists and chemists has generally eluded them, because their knowledge of their subjects is always less comprehensive than is the knowledge of inanimate things. Psychologists generally cannot predict the future behavior of an individual on the basis of past evidence with a high degree of accuracy. The same is true of sociologists and the behavior of populations.

1.3 LANGUAGE AND VAGUENESS

A second phenomenon imposing limits upon our desire for precision is that we use natural language in order to both describe and communicate knowledge. All of us have had the experience of misunderstandings that result from having used words in a different way than our conversation partner used them. Our understanding of word meanings carries with it the full texture of cultural and personal associations, so, even though we share core meanings and are thus able to communicate accurately to an acceptable degree most of the time, we generally cannot absolutely and precisely agree among ourselves on one single word meaning. To say it another way, natural language has, to a significant extent, the characteristic of *vagueness*.

An example, often cited in discussions about linguistic uncertainty, is the so-called *heap paradox*. We say that a *heap* is a collection of parts, such as a heap of grain or a heap of stones. If we remove one part, we will still be left with a heap, since no one individual's absence makes a big difference to the identity of our collection as a heap. Take away another part, and we still have a heap. After repeatedly removing individual parts, there will be a time, when we will be left with, say, two individuals: two rocks or two kernels of grain. Clearly, these do not constitute a heap any longer. So, our question is, at which point did our collection cease to be a heap? We cannot really identify the exact number of objects that must remain in the collection in order for it to qualify as a heap, because we would probably still be willing to call it a heap, even if we removed one more item. The significance of this is that any investigation employing a concept, such as "heap" will have to deal with vagueness.

Vagueness may also be the result of several users of a natural language accepting a slightly different meaning of a term. For example, a person accustomed to living in San Diego may have internalized a meaning for the term 'cloudy' that is different from the meaning accepted by a resident of Seattle. Both speakers of English would probably agree on some core meaning for 'cloudy': for one thing, neither would apply the term to a day in which no cloud appeared in the sky at all. However, the resident of Seattle would probably be more tolerant of clouds and be unwilling to call the sky cloudy if there were a cloud cover of 30 percent. For the resident of San Diego, however, a cloud cover of such a degree might be consistent with her concept of cloudiness.

Vagueness, in some contexts, presents a problem to scientists whose work requires great precision. Physicists need to agree on the meaning of expressions such as 'force,' 'space,' 'electron,' and so on. In order to achieve precision and avoid uncertainty and error, they stipulate—or prescribe—the meanings of their professional technical language. But having done so, they do not converse in the natural language alone: They have augmented it with a specially constructed language, an *artificial language* with precise and single-layered meanings.

Artificial languages of this sort exist in many disciplines, in mathematics, for example, or in engineering, because artificial languages can be represented in quantitative, computational ways. However, even though these languages are useful, they generally do not convey all the ideas such a language is expected to articulate. This is partly because what is gained by making the language precise is lost in its ability to describe the full extent of our experience, which is multi-layered. Hence, we need a quantitative method for taking into account vagueness and making the most of its many layers of meaning for the sake of dealing with complex problems. Indeed, such a method is particularly useful in those disciplines in which measurement is often imprecise, such as in some areas of psychology, biology, and sociology.

1.4 THE EMERGENCE OF FUZZY SET THEORY

From the beginning of modern science until the end of the nineteenth century, uncertainty was generally viewed as undesirable in science and the idea was to avoid it. This attitude gradually changed with the emergence of statistical mechanics at the beginning of the twentieth century. To deal with the unmanageable complexity of mechanical processes on the molecular level, statistical mechanics resorted to the use of statistical averages and probability theory. After its acceptance in statistical mechanics, probability theory has been successfully applied in many other areas of science. However, in spite of its success, probability theory is not capable of capturing uncertainty in all its manifestations. In particular, it is not capable of capturing uncertainty resulting from the vagueness of linguistic terms in natural language. These limitations, which are described at greater length in Sec. 1.5, are part of the reason why a new uncertainty theory, able to deal with imprecision and vagueness, was conceived.

We may identify the origin of the new theory of uncertainty, distinct from the concept of probability, as the publication of a seminal paper by Lotfi A. Zadeh in 1965.[1] Zadeh was interested in the problems of complex systems and how to represent them using simple models. Traditional mathematical tools proved to be unsatisfactory for this purpose. In his paper, Zadeh introduced the concept of a *fuzzy set*, a set whose boundary is not sharp, or precise. This concept contrasts with the classical concept of a set, recently called a *crisp set*, whose boundary is required to be precise. That is, a crisp set is a collection of things for which it is known whether any given thing is inside it or not. An example of a crisp set is the collection of coin types that constituted legal American coinage in 1994: penny, nickel, dime, quarter, half-dollar, and "Susan B. Anthony" dollar. So, the concept "legal U. S. coin type" has as its counterpart the set named LEGAL U. S. COIN TYPE. With respect to any recognizable, undamaged coin in our pocket, it will either be one of the coin types listed above or not. If we ask the question, "Is this particular coin in my pocket one of the six types of U. S. coins?" the answer will either be definitely *yes* or definitely *no*. The membership of that coin in the set of U. S. coin types will, therefore, *not* be a matter of degree. That is, the set has a sharp boundary, as illustrated in Fig. 1.1. The six types of U.S. coins are inside the boundary. Any other coin type (French franc, English pound, etc.) is outside the boundary.

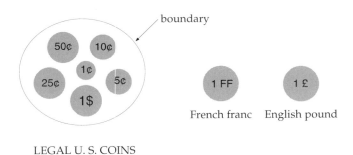

LEGAL U. S. COINS

Figure 1.1   Example of a crisp set.

1. L. A. Zadeh, "Fuzzy Sets," *Information and Control* **8**(3), 1965, pp. 338–353.

Contrary to classical crisp sets, fuzzy sets do not have sharp boundaries. That is, being a member of a fuzzy set is not a simple matter of being definitely *in* or definitely *out*: A member may be inside the set to a *greater or lesser degree*. For example, the concept "warm" includes a wide range of temperatures, as measured on some scale. This concept may be regarded as the name of a set, WARM, of many different temperature readings. In a given context, say a weather forecast at some place and time, we might consider several possible temperature readings, such as 65°, 75°, 85°, and 95°, and ask to what extent each reading is compatible with the concept "warm." Depending on our point of view and specifics of the context (place, season, day, night, etc.), we might say that the reading of 95° definitely belongs inside the set WARM. It is likely that we would make the same judgment about the readings 75° and 85°, though this is not certain for every situation. However, the reading 65° would not appear to all of us to be clearly in this set: It is a borderline air temperature reading. For some of us, 65° is already a little chilly, especially if we are accustomed to a southerly climate. But for those from the upper midwestern United States, 65° is often the high temperature of the average summer day! Accordingly, one would say that the phrase 'is a warm reading' is at least somewhat true for 65°; certainly it is not clearly true and it is also not clearly false. Because 65° and other temperatures are in the set WARM to various degrees, the set is a fuzzy set. We can define this fuzzy set by assigning to each temperature a number between 0 and 1, which indicates the degree or grade of membership in the set. The assignment of 0 to a particular temperature means that this temperature definitely does not belong to the set; the assignment of 1 means that the temperature definitely does belong to the set.

The assignment of a grade of membership in the fuzzy set WARM to each considered temperature is called a *membership grade function* of this fuzzy set. An example of this function, reflecting a particular context, is shown in Fig. 1.2.

We can see that each fuzzy set is uniquely determined by a particular membership grade function, which assigns to each object of interest its grade of membership in the set. Although it is not necessary, it is convenient to express the grade of membership by numbers between 0 and 1. Membership functions are discussed in detail in Chapter 4.

Because membership in a fuzzy set is not a matter of affirmation or denial, such as *yes* or *no, true* or *false*, 1 or 0, certain traditionally accepted rules governing reasoning are not accepted as rules in

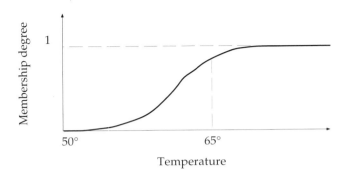

Figure 1.2   Possible membership function of fuzzy set WARM.

the new theory. For example, if we ask, whether an air temperature of 65° Fahrenheit is a member of the set of warm temperatures (WARM) or of the set of cold temperatures (COLD), then we are willing to say that it could belong to both sets with degrees that depend on one's physical constitution or context. Accordingly, the traditional rule that the same thing cannot be in a set and its complement at the same time does not operate in this situation. By contrast, if we refer back to the crisp set LEGAL U. S. COIN, my French franc coin will definitely be either in this set or in the set of coin types that are not legal U. S. coin types in 1994, *but not in both*. Consequently, the new theory of uncertainty—fuzzy set theory—must take into account that the rule of non-contradiction will not work in it. This is a consequence of accepting certain sets such as WARM and COLD, as fuzzy sets, whose boundaries are not sharp.

Zadeh's proposal of a theory of fuzzy sets did not, at first, attract a wide following, but he and a small group of other researchers continued to develop it over the decade that followed. Still, even though development of the theory progressed slowly, some important contributions were made during the 1970s. For example, the important idea of fuzzy control emerged in the early 1970s and the first fuzzy controllers were built later in the decade. These controllers are based on inference rules stated in natural language and represented by fuzzy sets. Several important ideas regarding the use of fuzzy sets in pattern recognition and clustering also emerged in the 1970s. During this time, too, the International Fuzzy Systems Association (IFSA) was formed; moreover, the official journal of this association, *Fuzzy Sets and Systems*, began publication.

The decade of the 1980s witnessed not only the increasing pace of theoretical development of fuzzy set theory, but it also ushered in a time of practical applications of the theoretical discoveries. Synthesizing the results of researchers in many countries, manufacturers—especially those in Japan—have brought products into the consumer market whose functional designs are based on fuzzy set theory. This is a trend which is continuing, so that in the 1990s there are, for example, automobiles whose transmissions, anti-skid brakes, and other functions are controlled by fuzzy logic circuits.

Interest in such engineering applications is continuing to build, but it is an interest not confined to manufacturing. The world of finance and investment has also discovered the utility of applying fuzzy theory to its decision support activities. For example, stock market analysts and investment managers have found that fuzzy set theory enables them to understand the underlying regularities of stock price fluctuations and to predict future behavior of stocks more accurately than with traditional analytic methods. Successful applications of fuzzy set theory have also been established in such areas as expert systems, database and information retrieval systems, pattern recognition and clustering, signal and image processing, speech recognition, risk analysis, robotics, medicine, psychology, chemistry, ecology, and economics. A survey of applications of fuzzy set theory is covered in Chapter 10.

Today, we are experiencing the emergence of interdisciplinary projects in which fuzzy set theory plays an essential role. This is especially true in projects involving the design of intelligent systems. These are human-made systems that are capable of achieving highly complex tasks in a humanlike, intelligent way. In these projects, fuzzy set theory is often combined with neural network computation, genetic algorithms, evolutionary computation, and other advanced methodological tools.

1.5 FUZZY SET THEORY VERSUS PROBABILITY THEORY

Probability theory is one of the most important traditional theories describing and systematizing the phenomenon of uncertainty, and, thus, it is sometimes thought that it can be used in every situation of uncertainty. According to this view, we do not require the theory of fuzzy sets, since probability theory is thought to be sufficient for dealing with all kinds of uncertainty.

This view is ill-conceived since probability and fuzzy sets describe different kinds of uncertainty. One way to describe the difference is to say that probability theory deals with the *expectation* of a future event, based on something known now. For example, we might be interested in the probability that the next person walking into our classroom is tall, where our concept of tallness is one that matches the distribution of heights over the total U. S. population. If we know that we are in a classroom restricted to basketball players, we will have a strong expectation that the next person entering the room will be tall. On the other hand, if we are in a classroom restricted to equestrians, our expectation will be much lower. Hence, our sense of uncertainty here revolves around making a prediction about an event.

The sense of uncertainty represented by fuzziness, however, is not the uncertainty of expectation: It is the uncertainty resulting from the imprecision of meaning of a concept expressed by a linguistic term in natural language, such as *tall, warm, very warm, rapidly increasing*, and the like. Suppose we are in a classroom of basketball players, and a player who is six feet, one inch tall enters the room. At this point, would it make sense to ask, "What is the probability that this player is tall?" Now that he is in the room, we no longer find ourselves in a situation of expectation; we know for sure that he has this height and that his height is consistent to a great extent with the ordinary meaning of the term 'tall.' The more appropriate question is, "To what extent is it correct to say that this player is tall?" If we compare him to the total U. S. population, we would answer, "it is correct to a great extent." If, instead, we compare him to the basketball team, we might be able to answer only that it is correct to a small extent that he is tall, if the average height of basketball players is six feet, eight inches. Expectation plays no role here; we are describing the compatibility of a particular characteristic (the height) of an individual with a given imprecise concept (tallness).

Another important way to draw the distinction between probability and fuzzy sets is to point out that probability is the theory of *random* events. For example, the appearance of a "head" on a coin is a random coin tossing event. Probability theory is concerned with the likelihood of relevant events. Fuzzy set theory, on the other hand, is not concerned with events at all. It is concerned with concepts, such as "tall" or "warm," and whether or not an individual, such as a particular person, or a state, such as a particular temperature reading, matches the meaning of the concept in question.

There are also many situations that exhibit both kinds of uncertainty. For example, a weather forecaster might warn us that it is highly probable that it will be cloudy tomorrow. In this example, not only is "cloudy" a fuzzy concept, but "highly probable" is also a concept that involves both randomness and fuzziness. At the bottom of this issue is that probability theory and fuzzy set theory are useful for characterizing different kinds of uncertainty. Hence, rather than competing with each other, they complement each other and may often be combined.

EXERCISES

1.1 Describe paradoxes analogous to the heap paradox. For example, the paradoxes of bald men, tall persons, expensive cars, and the like.

1.2 Using the membership function in Fig. 1.2 as a reference, define comparable membership functions of fuzzy sets COLD, VERY COLD, VERY WARM, and the like.

1.3 Define membership functions that express the meaning you associate with various concepts frequently encountered in daily life, such as *low interest rate, high inflation, very high tuition, acceptable salary, short driving distance, expensive dinner, old wine, highly educated person, modest academic performance, large university*, and the like.

1.4 Explain the differences between fuzziness and randomness; find some examples dealing with fuzziness, randomness, and both.

2

CLASSICAL LOGIC

2.1 INTRODUCTION

Classical logic is the study of the *forms of correct reasoning* and is, therefore, also called *formal logic*. It is not the study of psychological processes involved in reasoning, or the study of whether or not our conclusions are made true by any facts in the world. Rather, its focus is on *abstract, basic patterns of reasoning*. By *correct* reasoning we mean the kind of reasoning which ensures that true conclusions follow from true premises or perfect evidence. Using correct forms of reasoning, we are able to trust our conclusions when they are based on true premises.

All major philosophical systems, Eastern and Western, have developed at least one system of formal logic. In Western intellectual history, we generally consider the Greek philosopher Aristotle to have provided the first systematic account of correct forms of reasoning. His system prevailed as the dominant system of formal logic until the nineteenth century. Since then, his system has been augmented by other means of representing logical forms, namely those in which natural language is represented by a small set of symbols capable of reflecting the fundamental structure of reasoning with full precision, which is the aim of classical logic. This kind of logic is called *symbolic logic* and divides into *propositional logic* and *predicate logic*.

2.2 PROPOSITIONAL LOGIC

Propositional logic is a formal system for representing knowledge in terms of declarative sentences that express propositions, using a method of letters and symbols which stand for such propositions and the logical connections between them. When we employ this method, we are interested in the inference patterns of our reasoning, and so we are interested in the structure of inferences. Inferences are just lists of declarative propositions, one of which is the conclusion and the rest of which are the reasons, or premises, supporting that conclusion. Inference patterns may divide into two broad types: inductive patterns and deductive patterns. Inductive patterns are not part of the subject matter of this book and will be described only for the purpose of drawing the distinction between them and deductive patterns, as well as the distinction between *certainty* and *uncertainty*.

Forms of Reasoning

Two broad types of reasoning may be distinguished: *inductive* reasoning and *deductive* reasoning. Their distinction is based on the relationship between the reasons presented in support of a conclusion and the conclusion itself. That is, the distinction relates to the strength of support the reasons—or *premises*—of an inference lend to the conclusion. Consider how a student might reason, when planning her day's classes:

> If today is Monday, then my logic class meets at noon.
> Today is Monday.
>
> ─────────────────────────────────
>
> So, my logic class meets at noon.

Let us suppose that she is right about her logic class schedule and that the first premise is true. Suppose also that she is correct about the day of the week. We can see that her conclusion is true beyond a shadow of a doubt. Indeed, we cannot think of any circumstance under which the two premises could be true and the conclusion false. We will soon learn that the *certainty* of her inference is justified by the interplay between the correct form of this inference and the truth of the premises. But for now, we should just realize that in this correctly structured reasoning, once the premises are accepted as true, there is no possibility that anything could make the conclusion false. This is the hallmark of correct deductive reasoning:

> *If an inference is a correct deductive inference, then it is impossible for its premises to be true and its conclusion to be false. Thus, the relationship between premises and conclusion is one of certainty.*

By contrast, consider this line of reasoning about observing the behavior of meteors:

> Meteor 1 disintegrated upon entering the Earth's atmosphere.
> Meteor 2 disintegrated upon entering the Earth's atmosphere.
> Meteor 3 disintegrated upon entering the Earth's atmosphere
>
> $\vdots$
>
> Meteor 50 disintegrated upon entering the Earth's atmosphere.
> _____
> Thus, all meteors disintegrate upon entering the Earth's atmosphere.

Here we have a conclusion which makes a *general* claim about all meteors; it does not make a claim simply about the observed objects. However, the conclusion is supported by a significant number of examples. We would be prepared to say that the observer has fairly good evidence for his conclusion, and we would agree that his reasoning is sound. However, it is clear that the relationship between the set of 50 premises and the conclusion they support is not one of certainty. It is easy to imagine that the observer happened to witness the behavior of a particular cluster of meteors that all had the same physical composition. Being physically similar, they reacted to the friction of the atmosphere in the same way. However, we know that meteors may be composed of many different substances and that many of these would react—and have reacted—differently to their atmospheric encounter. This means that although the observer was correct about each of his 50 observations, it is possible for the conclusion to be false.

The lesson of this example is that there are many complex cases of reasoning whose conclusions are not supported by their premises with certainty; they are supported with more or less likelihood, and thus there is always some element of uncertainty in reasoning which uses this kind of inferential structure. Reasoning of this type is called *inductive* and is studied in specialized courses concerned with the rules of induction. We, however, restrict ourselves to exploring the most important features of *classical deductive logic*.

Form, Validity, and Truth

The sample deductive inference above is clearly correct. We know that once the premises are accepted as true, the conclusion must be true as well. The reason that this is so is that the inference has a certain kind of structure which ensures that true premises produce a true conclusion. If letters M, L denote the propositions "today is Monday" and "my logic class meets at noon," respectively, then we would have the following schematic representation:

If M, then L

M

_____

Therefore, L

When we depict an inference in this way, we want to convey the idea that if both premises are true, then the conclusion must also be true.

Another way to represent the inference is to allow the arrow ⇒ to stand for the relation *if,...then* and for the symbol ∴ to stand for "therefore." Further, we can recognize that there is an underlying, general structure inherent in this inference. We may represent it by allowing general place holders to replace our propositional letters:

The symbols □ and ○ in this structure may be seen as "place holders" for propositions. Finally, we recognize that any inference which shares this general structure produces a true conclusion when both of the premises are true. This means that the structure has the characteristic of *validity*. That is, an inference that is valid is a correct deductive inference.

The Structure of Propositional Logic

As already mentioned, the basis of classical propositional logic is that inferences are composed of declarative statements, or propositions, which may be true or false. Recall that we are assuming that the meanings of these declarative propositions are crisply defined and that their truth values are also crisp. Let us now stipulate that the building blocks of inferences be simple, affirmative, declarative propositions which may be combined to form complex propositions. By a *simple declarative proposition*, we mean the same thing as is meant by a grammatically simple proposition, namely a proposition that does not contain any other proposition as a component. Such simple propositions are also called *atomic* propositions. A proposition is *affirmative* when it contains no negating words or prefixes. Examples of such atomic propositions are

A dog has four legs.

Tomorrow is Sunday.

Brazil has won the World Cup.

The solar system contains nine planets.

Let *simple propositions* be represented by single, lowercase letters, such as *p, q, r,* or *s*. When the meaning of each such letter varies with the specific proposition which it represents at any particular time, then the letters assume the role of *propositional variables*.

Complex propositions are constructed by applying certain logical operations to atomic propositions. For example, we may apply the operation of conjunction, or simultaneity, to the first two propositions in our list above and get the complex proposition

A dog has four legs and tomorrow is Sunday.

If we let the propositional variable p replace 'A dog has four legs' and the propositional variable q replace 'tomorrow is Sunday,' then this complex proposition would be written

$$p \text{ and } q$$

This operation and several others are considered in the next section.

Logic Operations

Let us consider five logic operations to form our symbolic language: negation, conjunction, disjunction, implication, and equivalence. Although other logic operations are possible, these five operations, which are sufficient to describe any complex proposition, are the most commonly employed operations in formal logic. Table 2.1 assigns a symbolic representation to each of these operations.

TABLE 2.1   LOGICAL CONNECTIVES

Operation	Symbol	English Examples
negation	$\neg$	not; it is not the case that, un-; im-; in-
conjunction	$\wedge$	and; but; however
disjunction	$\vee$	or; unless
implication	$\Rightarrow$	if, ... then; implies; only if
equivalence	$\Leftrightarrow$	if and only if; when and only when

Negation

The simplest operation that we apply to propositions is negation. For example, the negation of the proposition p, 'a dog has four legs,' is $\neg p$, 'a dog does NOT have four legs,' or 'it is not the case that a dog has four legs.' Similarly, the affirmative proposition q embedded in the negated proposition $\neg q$, 'Elvis is immortal,' is 'Elvis is mortal,' where the prefix "im-" represents its negative particle. In summary:

English Sentence	Symbolic Representation
A dog does NOT have four legs	$\neg p$
Elvis is immortal	$\neg q$

Because negation involves only one proposition, it is called a *unary operation* and the symbol that represents this operation is called a *unary operator*. This operator is not only useful for representing negations in natural-language propositions, but it also reveals its behavior with respect to the ways it can be true or false.

The negation of a proposition has the opposite truth value of that proposition. Accordingly, if a proposition is true, its negation is false, and vice versa. We can represent this behavior by a so-called *truth table* in which T stands for "true" and F stands for "false." The propositional variable p is a place holder for any proposition:

p	$\neg p$
F	T
T	F

Though this representation is commonly used in texts on philosophical logic, we employ an alternative representation, also frequently used in the literature, in which TRUE is replaced by the number 1 and FALSE by the number 0:

p	$\neg p$
0	1
1	0

Observe that the numbers 0 and 1 are used here purely as convenient symbols, particularly for computer representation, but they have no numerical significance.

The column of truth values under the proposition variable p is called the *basis column*, because the values depicted in it are the basis for determining the truth values of the proposition $\neg p$ for each of the two values that the proposition p can possibly have.

An alternative way to describe the truth-functional behavior of negation is to give it a one-line description. We adopt the symbolic expression

$$|p|$$

to mean

'the truth value of the proposition denoted by p'

Accordingly, the description of negation is

$$|\neg p| = 1 - |p|$$

We can see, for example, that if $|p|$ has the value 1 (True), then $|\neg p|$ has the value 1–1 = 0 (False).

Conjunction

A conjunction is formed by connecting two propositions with a conjoining logical word, such as 'and.' Because it involves two propositions (conjuncts), it is called a *binary operation;* its symbol is a connective, because it always connects two propositions. Examples of English language conjunctions include propositions such as these:

Today is Wednesday and tomorrow is Thursday.

Binghamton is a town and New York is a state.

John graduated summa cum laude; moreover, he received a poetry prize.

The truth table of conjunction, given in Table 2.2, tells us how conjunction behaves with respect to the truth values of the propositions involved. This truth table shows that a conjunction may be true only when both conjuncts are true at the same time; otherwise, the conjunction is false. Hence, this operation is very restrictive. The truth table further makes it clear that conjunctions, like negations, are *truth functional*. This means that each provides us with a pattern containing the relevant connective and some place holders for possible propositions that have truth values. We are able to "fill up" the place holders with truth values, and then the connective gives us a unique assignment of the overall truth value of the conjunction. Thus, the truth value of a complex proposition is a function of the individual truth values of the constituent propositions.

As in the case of negation, we can provide a one-line description of the truth-functional behavior of conjunction. This description expresses the fact that the truth value of a conjunction is always equal to the *minimum* (min) value had by its conjuncts:

$$|p \wedge q| = \min [|p|, |q|]$$

TABLE 2.2   CONJUNCTION

p	q	$p \wedge q$
0	0	0
0	1	0
1	0	0
1	1	1

This expression is read "the truth value of a conjunction is the minimum of the truth values of p and q." Thus, if $|p| = 1$ and $|q| = 0$, then the truth value of their conjunction is the minimum value between them, namely 0. Observe, however, that alternative descriptions of conjunction are possible, for example,

$$|p \wedge q| = |p| \cdot |q|$$

or

$$|p \wedge q| = \max [0, |p| + |q| - 1]$$

While these descriptions are equivalent in classical logic, they are distinct in fuzzy logic, as explained later in the text.

Disjunction

Disjunctions are binary logic operations formed by connecting two propositions with a disjoining logical word, such as 'or.' Among the many English language examples of disjunction are propositions such as

> Jupiter has ten moons or it is a massive planet.
> A double major or a high grade point average is impressive for
>   law school.
> John plays football or basketball.

Disjunction is true if at least one of the two propositions (disjuncts) involved is true; its truth table is given in Table 2.3.

The one-line description of the truth-functional behavior of disjunctions says that their truth value is the *maximum* (max) truth value of its disjuncts:

$$|p \vee q| = \max [|p|, |q|]$$

TABLE 2.3   DISJUNCTION

p	q	$p \vee q$
0	0	0
0	1	1
1	0	1
1	1	1

Accordingly, if $|p| = 0$ and $|q| = 1$, then their disjunction is the maximum of 0 and 1, namely, 1 (True). Again, other descriptions are possible, for example,

$$|p \vee q| = \min\,[1, |p| + |q|]$$

which are equivalent in classical logic, but not in fuzzy logic.

Disjunction, whose linguistic meaning is "A or B" or, alternatively, "either A or B or both," must be distinguished from another logic connective whose meaning is "either A or B but not both." The latter is sometimes called an *exclusive disjunction*, but it can also be viewed as logic nonequivalence, as mentioned later.

Implication

Implications are also known as *conditional* propositions and are usually, but not always, encountered in English in the form of "if,...then" propositions. Examples of implications include propositions such as

> If I meet the university distribution requirements, then I can graduate.
> On condition that the vehicle is maintained properly, it will run for at least 200,000 miles.
> We will have a celebration only if we are victorious.

The first—"if"—part of an implication is called the *antecedent* and the second—"then"—part is called the *consequent*. This connective requires for its truth value that no implication may be true if its antecedent is true, but its consequent is false. Under all other conditions, implications are accepted as true. Table 2.4 represents these principles. This truth table tells us that if we represent an implication relation between two propositions, then we certainly want to call the implication "true," if both antecedent and consequent are true, as in

> If today is Sunday, then the *New York Times* has more than 20 pages.

TABLE 2.4   IMPLICATION

p	q	$p \Rightarrow q$
0	0	1
0	1	1
1	0	0
1	1	1

Further, we want to call an implication "false," if we are able to make the antecedent true, but the consequent turns out to be false, as occurs in this proposition:

> If the moon is the Earth's satellite, then it is composed of water.

On the other hand, any implication with a false antecedent is considered to be true. The reason is that we have not had the opportunity to prove decisively the implication false, since the condition in the antecedent has not been established. We can see this in the following example:

> If this is the year 2057, then everyone has a solar-powered housekeeping robot.

Since it is not yet the year 2057, we do not have the opportunity to test the truth of the consequent, and thus have no definite reason to say that it is false. Hence, in our classical, two-valued logic, we are required to consider this implication to be provisionally true.

Two possible one-line descriptions of implication, which show its truth-functional character are:

$$|p \Rightarrow q| = \min(1, 1 + |q| - |p|)$$
$$|p \Rightarrow q| = 1 - |p|(1 - |q|)$$

These descriptions are again equivalent in classical logic, but not in fuzzy logic. We will see this in later chapters.

Observe that implication is an *asymmetric* logic connective in the sense that

$$\lvert p \Rightarrow q \rvert \neq \lvert q \Rightarrow p \rvert$$

This is contrary to conjunction and disjunction, which are *symmetric*.

Equivalence

Finally, let us consider our last binary connective, *equivalence*—also called *material equivalence* or *biconditional*. It is encountered in English most often in the logical phrase 'if and only if':

> You will become a star if and only if you have an enterprising agent.

Actually, a proposition of the form "*p* if and only if *q*" just asserts the same thing represented by the following conjunction:

$$\text{if } p, \text{ then } q, \text{ and if } q, \text{ then } p$$
$$(p \Rightarrow q) \wedge (q \Rightarrow p)$$

However, using a special equivalence symbol is much more economical than using the conjunction of two implications. Accordingly, we write the above propositional form in this way:

$$p \Leftrightarrow q$$

The truth table for equivalence concentrates entirely on the question of whether or not two propositions have the same truth values. That is, we are interested in knowing if two propositions are true at the same time or false at the same time. Accordingly, the characteristic truth table for "if and only if" is as given in Table 2.5.

We can see that we truly assert in rows 1 and 4 that the two propositions composing the equivalence are true at the same time in one case and false at the same time in the other. In the two cases in the middle, however, it is false that both *p* and *q* have the same truth values. Another way to describe this truth table is to say that the equivalence asserts that two propositions are *equivalent with respect*

TABLE 2.5   EQUIVALENCE

p	q	$p \Leftrightarrow q$
0	0	1
0	1	0
1	0	0
1	1	1

to truth value. When $p \Leftrightarrow q$ is true, it also means that p is a necessary and sufficient condition for q, and vice versa.

A possible one-line description of the truth-functional behavior of equivalence is

$$|p \Leftrightarrow q| = |p| \cdot |q| + |\neg p| \cdot |\neg q|$$

The five logical connectives—negation, conjunction, disjunction, implication, and equivalence—are sufficient to capture the underlying logical structure of all complex propositions, and thus we say that they are a *complete* set of connectives. In fact, various subsets of these connectives are also complete in this sense, for example, negation with conjunction, negation with disjunction, negation with implication, and so on. Any complete set of logic connectives that is chosen to represent complex propositions is usually referred to as a *set of logic primitives.*

Truth Values of Complex Propositions

It is easy to see that each of the five logic connectives discussed in previous sections allows us to represent various combined propositions: Some contain conjunctions, others contain several implications, and so on. However, we are also able to represent much more complex propositions containing two or more different connectives. Consider the following proposition:

Either Joe or Bill or both will play badly at the tournament, but if Joe plays badly, then there will be no splashy victory party.

This proposition is represented by the symbolic expression

$$(p \vee q) \wedge (p \Rightarrow \neg r)$$

where

> p: Joe plays badly at the tournament.
> q: Bill plays badly at the tournament.
> r: There will be a splashy victory party.

The main connective of this proposition is the conjunction, each of whose conjuncts has some additional structure. We can determine the main connective by noting that the pairs of parentheses on either side of the conjunction symbol leave it outside the scope of these two "punctuation" symbols. This shows that we must use parentheses or some similar notation to avoid a succession of two or more connectives, none of which inherently has precedence over the others. Parentheses establish which of the operations must be performed first, which second, and so on. The rule is that the innermost connective(s) must be computed first; then we work outward to the main connective. If we arbitrarily assign the value 0 to p and r, and the value 1 to q, then we work out the overall truth value of the proposition according to the schema in Fig. 2.1.

Just as we were able to represent the truth-functional behavior of our basic logic expressions, so we are able to represent the truth-functional behavior of this more complicated proposition. In order to represent this behavior, we construct a truth table containing a number of rows representing all possible combinations of truth values. For n logic variables, the two truth values of each variable are combined with the truth values of other variables. This results in 2^n combinations, each representing a row in the truth table.

In our case, since $n=3$, we have eight rows, as shown in Table 2.6. The columns identified by the labels a, b, c, d are the truth values

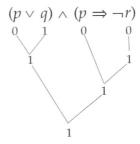

Figure 2.1   Evaluation of a logic expression.

of the constituent parts of the entire proposition. Clearly, a, b, c must be calculated before d, and a must be calculated before b.

TABLE 2.6 TRUTH TABLE FOR A COMPLEX PROPOSITION

p	q	r	$(p \lor q) \land (p \Rightarrow \neg r)$			
0	0	0	0	**0**	1	1
0	0	1	0	**0**	1	0
0	1	0	1	**1**	1	1
0	1	1	1	**1**	1	0
1	0	0	1	**1**	1	1
1	0	1	1	**0**	0	0
1	1	0	1	**1**	1	1
1	1	1	1	**0**	0	0
			c	d	b	a

Propositional Forms

Consider again the logic expression

$$(p \lor q) \land (p \Rightarrow \neg r)$$

When symbols p, q, r are considered as logic *variables*, the expression is nothing more than a *general propositional form* that may be shared by many actual, natural-language propositions. For the rest of this topic, we deal only with general propositional forms.

Contingent Propositions

The truth table for the propositional form above demonstrates that any proposition in this form is sometimes true and sometimes false. Each row establishes the possible truth-value combinations of the three logic variables. Each of these variables, when replaced by a natural-language proposition making an assertion, will have some truth value assigned to it in a particular row and thus reflects some aspect of external reality. Other rows reflect reality differently, and so the complex proposition may have a different truth value. Another way to describe the eight rows is to say that each provides an *interpretation* of the variables.

Our complex propositional form is the form of a proposition whose overall truth value depends on the way each of the eight possible truth-value combinations appears. When the facts are one way, the proposition turns out to be true; when the facts change, it may turn out to be false. Thus we say that the truth value of our proposition at each row is *contingent* upon the truth values generated by each assignment of truth values. The propositional form itself is, therefore, said to be a *contingent* propositional form. Since propositions in this form make assertions about something in the world, they are also called *empirical propositions*, propositions whose truth value can be known only by consulting our experience of reality.

Contradictions

Contradictions are complex propositions whose truth tables show them to be false under all assignments of truth values to their constituent atomic parts. The most paradigmatic form of a contradiction is the conjunction $p \land \neg p$, which says that the same proposition is affirmed and denied at the same time. Also called the *law of contradiction*, the truth table below asserts that a contradiction cannot be true under any interpretation—truth value—of its constituent propositions:

p	$p \land \neg p$
0	0
1	0

We can see that we can characterize a contradiction as a proposition form whose truth value is independent of the truth values of its components: No facts in the world will ever change the truth value of a contradiction. Hence, its truth value does not depend on anything in the world; it is the result entirely of the structure of the proposition.

Tautologies

Tautologies are also often called *necessarily true propositions* or *analytic propositions*. As one of its names suggests, tautologies are *true under all assignments of truth values* to their component propositions: No facts can ever cause them to become false, and thus they are true entirely because of their internal structure. The most basic form of a tautology is the propositional form $p \vee \neg p$, which is usually called the *law of the excluded middle*:

p	$p \vee \neg p$
0	1
1	1

Among the most frequently encountered tautologies are those in which the main connective is an implication. These are conditional propositions in which it is impossible to find a case where the antecedent is true, but the consequent false. Another way to say this is that tautologous conditional propositions are such that if their antecedents are true, their consequents must be true. For example, the conditional proposition in Table 2.7 is a tautology and is called a *logical implication*, or *entailment*. We show later that propositional forms such as this one describe valid logical inferences.

TABLE 2.7   A TAUTOLOGY: LOGICAL IMPLICATION

p	q	\[$(p \Rightarrow q)$	$\wedge$	p\] $\Rightarrow q$		
0	0		1	0	1	
0	1		1	0	1	
1	0		0	0	1	
1	1		1	1	1	
			a	b	c	

Another important class of tautologies consists of necessarily true equivalences. We saw earlier that material equivalences are sometimes true and sometimes false, so they are contingent. But some equivalences are never false, as Table 2.8 shows.

TABLE 2.8 A TAUTOLOGY: LOGICAL EQUIVALENCE

p	q	$\neg(p \wedge q) \Leftrightarrow (\neg p \vee \neg q)$
0	0	1 0 **1** 1 1 1
0	1	1 0 **1** 1 1 0
1	0	1 0 **1** 0 1 1
1	1	0 1 **1** 0 0 0
		f e g b c a

This equivalence truth table tells us that whatever makes one side of the equivalence true also makes the other side true; and whatever makes one side false, also makes the other side false. This is because the two sides have the same meaning and are, therefore, interchangeable. Such propositions are called *logical equivalences*. In effect, logical equivalences can be thought of as definitions, or rules for using language. Table 2.8, for example, tells us that we may replace any expression of the form $\neg(p \wedge q)$ by an expression of the form $(\neg p \vee \neg q)$; the two expressions are synonymous and will preserve the truth value of any larger proposition in which they are interchanged. A later section presents several so-called rules of replacement in the form of logical equivalences which allow us to make synonymous substitutions of various expressions.

Logic Functions

We have seen several examples illustrating the truth-functional character of complex propositions: The truth value each kind of propositional form produces after each propositional variable—a placeholder—has been "given" a truth value is a *function* of that form and the actual values filling up the propositional placeholders. We saw that tautologies, for example, are produced by propositional functions in such a way that only the value "true," or 1, results under the main connective, no matter which truth values are given to the function. Just the opposite is the case for contradictions. Contingent propositional functions, on the other hand, are sensitive to the truth values given to them: They can produce a variety of different combinations of "true" and "false" under their main connectives.

It is possible to systematize these different kinds of logic functions by representing all of the possible patterns of "true" and "false." A *logic function* is a scheme by which a truth value of a propositional variable (called an *output variable*) is uniquely assigned to each combination of truth values of other variables (called *input variables*). For n input variables, there are 2^n combinations of truth values. Each logic function assigns one of two possible truth values to each of these combinations. The total number of distinct assignments, each representing one logic function of n input variables, is thus

$$\underbrace{2 \times 2 \times \ldots \times 2}_{2^n \text{-times}} = 2^{2^n}$$

For example, all logic functions of two input variables are shown in Table 2.9, where p, q denote input variables and r_1, r_2,..., r_{16} denote output variables of the individual functions. In Table 2.10, we list names that are often used for these functions in the literature and give typical examples of logic expressions that represent these functions. Observe that some of these functions are degenerate in the sense that the associated output variables either depend on only one of the input variables (assertions, negations) or are constant (tautology, contradiction).

TABLE 2.9   LOGIC FUNCTIONS FOR TWO VARIABLES: TRUTH VALUES

p	q	r_1	r_2	r_3	r_4	r_5	r_6	r_7	r_8	r_9	r_{10}	r_{11}	r_{12}	r_{13}	r_{14}	r_{15}	r_{16}
0	0	1	0	1	0	1	0	1	0	1	0	1	0	1	0	1	0
0	1	1	1	0	0	1	1	0	0	1	1	0	0	1	1	0	0
1	0	1	1	1	1	0	0	0	0	1	1	1	1	0	0	0	0
1	1	1	1	1	1	1	1	1	1	0	0	0	0	0	0	0	0

Inference Form and Validity

Truth tables may be used not only to determine the truth-functional character of propositions, but they are also helpful in allowing us to distinguish between valid and invalid inferences. We recall that the definition of 'validity' says that no valid inference may have all true premises followed by a false conclusion. We can use this definition to establish a truth-table test for validity.

TABLE 2.10   LOGIC FUNCTIONS FOR TWO VARIABLES: EXAMPLES

Function	Name	Expression Examples
r_1	tautology	$[(p \Rightarrow q) \wedge p] \Rightarrow q,$
r_2	disjunction	$p \vee q, \; \neg(\neg p \wedge \neg q)$
r_3	implication	$q \Rightarrow p, \; \neg q \vee p$
r_4	assertion	$p \vee (p \wedge q)$
r_5	implication	$p \Rightarrow q, \; \neg p \vee q$
r_6	assertion	$q \vee (p \wedge q)$
r_7	equivalence	$p \Leftrightarrow q, \; (p \wedge q) \vee (\neg p \wedge \neg q)$
r_8	conjunction	$p \wedge q$
r_9	not both	$\neg(p \wedge q)$
r_{10}	nonequivalence	$(p \vee q) \wedge \neg(p \wedge q)$
r_{11}	negation	$\neg[q \vee (p \wedge q)], \; \neg q$
r_{12}	inhibition	$p \wedge \neg q, \neg(p \Rightarrow q)$
r_{13}	negation	$\neg[p \vee (p \wedge q)], \; \neg p$
r_{14}	inhibition	$q \wedge \neg p, \; \neg(q \Rightarrow p)$
r_{15}	neither–nor	$\neg(p \vee q)$
r_{16}	contradiction	$(p \wedge \neg p) \wedge (q \wedge \neg q)$

Consider the following example:

Premise 1 $(p \Rightarrow q) \wedge (r \Rightarrow s)$
Premise 2 $p \vee r$
_____

Conclusion $q \vee s$

To test this inference, we form a truth table, Table 2.11, in which the two premises and the conclusion are all lined up at the top of the table, so we can scan across all three propositions simultaneously in each row. This truth table tells us that there is no possible combination of truth values for which all of the premises in this inference form are true, while the conclusion is false. Hence, this inference form is valid.

An example of an invalid form is

Premise 1 $p \Rightarrow q$
Premise 2 $\neg p$
_____

Conclusion $\neg q$

TABLE 2.11   TRUTH TABLE FOR A VALID INFERENCE FORM

p	q	r	s	$(p{\Rightarrow}q){\wedge}(r{\Rightarrow}s)$	$p{\vee}r$	$q{\vee}s$
0	0	0	0	**1**	**0**	**0**
0	0	0	1	1	0	1
0	0	1	0	**0**	**1**	**0**
0	0	1	1	1	1	1
0	1	0	0	1	0	1
0	1	0	1	1	0	1
0	1	1	0	0	1	1
0	1	1	1	1	1	1
1	0	0	0	**0**	**1**	**0**
1	0	0	1	0	1	1
1	0	1	0	**0**	**1**	**0**
1	0	1	1	0	1	1
1	1	0	0	1	1	1
1	1	0	1	1	1	1
1	1	1	0	0	1	1
1	1	1	1	1	1	1

This can be seen from the second row in Table 2.12.

TABLE 2.12   TRUTH TABLE FOR AN INVALID
INFERENCE FORM

p	q	$p{\Rightarrow}q$	$\neg p$	$\neg q$
0	0	1	1	1
0	1	**1**	**1**	**0**
1	0	0	0	1
1	1	1	0	0

Basic Inference Rules and Proofs

Applying truth tables to the problem of distinguishing valid from invalid inferences would sooner or later lead us to the realization that many inferences contain a large number of atomic propositions, with the effect that their truth tables have an overly large number of rows. Such truth tables might be too long ever to complete. Thus, on the one hand, the internal calculations of truth values are very simple, but—on the other hand—the size of the task is complex from the point of view of the resources necessary to generate a vast truth table. Clearly, we need to find a method in which complexity and simplicity are manifested differently. One such method is the so-called method of *natural deduction*.

Natural deduction makes it possible to use a small number of rules in a strategic way to demonstrate the validity of complex reasoning efficiently. Thus, we achieve relative simplicity in the task of demonstration. However, we lose the simplicity of the calculations which are done individually in truth tables. Instead, we must make use of a higher-level kind of thinking: the thinking involved in plotting a strategy to achieve a goal in some number of well-considered steps. Such strategic thinking has the characteristic of ingenuity, which is not required in the mechanical work of calculating truth values; hence, the method of natural deduction is sometimes also called the *method of ingenuity*.

We regard the method of natural deduction as a method in which we demonstrate, in a step-by-step fashion, that a conclusion must be true when a set of given premises is true. We perform this demonstration with the aid of a small set of *basic inference rules*.

Basic inference rules

In the same way in which complex propositions are composed of simple, atomic propositions, so complex, valid inferences are composed of simple, valid inferences. The demonstration of validity, therefore, involves the use of such simple inferences in the form of the basic inference rules in order to establish the truth-preserving connection between premise set and conclusion.

The most frequently used *basic inference rules* are those presented in Table 2.13. A test for the validity of each of these inference rules is that its premise set logically implies its conclusion. This means that if we form an implication by conjoining the premises of each two-step inference rule in the antecedent and place its conclusion into the consequent, then this implication will be a tautology; thus every inference rule may be expressed as a tautologous implication.

The central idea behind proving that a conclusion follows from a set of premises is to discover the ways in which the conclusion can be either found in or formed by the given premises. We accomplish this by

TABLE 2.13   BASIC INFERENCE FORMS

Conjunction (Conj.) 1. p 2. q $\quad\quad\overline{\quad\quad}$ $\therefore p \wedge q$	*Simplification (Simp.)* 1. $p \wedge q$ $\quad\quad\overline{\quad\quad}$ $\therefore p$
Addition (Add.) 1. p $\quad\quad\overline{\quad\quad}$ $\therefore p \vee q$	*Disjunctive Syllogism (DS)* 1. $p \vee q$ 2. $\neg p$ $\quad\quad\overline{\quad\quad}$ $\therefore q$
Modus Ponens (MP) 1. $p \Rightarrow q$ 2. p $\quad\quad\overline{\quad\quad}$ $\therefore \quad q$	*Modus Tollens (MT)* 1. $p \Rightarrow q$ 2. $\neg q$ $\quad\quad\overline{\quad\quad}$ $\therefore \quad \neg p$
Constructive Dilemma (CD) 1. $(p \Rightarrow q) \wedge (r \Rightarrow s)$ 2. $p \vee r$ $\quad\quad\overline{\quad\quad}$ $\therefore q \vee s$	*Destructive Dilemma (DD)* 1. $(p \Rightarrow q) \wedge (r \Rightarrow s)$ 2. $\neg q \vee \neg s$ $\quad\quad\overline{\quad\quad}$ $\therefore \quad \neg p \vee \neg r$
Hypothetical Syllogism (HS) 1. $p \Rightarrow q$ 2. $q \Rightarrow r$ $\quad\quad\overline{\quad\quad}$ $\therefore p \Rightarrow r$	*Absorption (Abs.)* 1. $p \Rightarrow q$ $\quad\quad\overline{\quad\quad}$ $\therefore p \Rightarrow (p \wedge q)$

performing the operations formalized in the basic inference rules on any one or two premises at a time, writing down an intermediate conclusion under the premise set and justifying this intermediate conclusion by reference to the numbers of the premises used and the name of the rule we applied. In this way, we arrive at the conclusion in the last step of every proof.

Consider the following example:

$(p_1)\, p \vee q$
$(p_2)\, \neg r \Rightarrow \neg p$ } premises given as true
$(p_3)\, r \Rightarrow s$
$(p_4)\, \neg s$

$(c)\ \therefore q$ conclusion to be established

$(r_1)\, \neg r \quad (p_3), (p_4)\ \text{MT}$ } results and reasons for new steps
$(r_2)\, \neg p \quad (p_2), (r_1)\ \text{MP}$ (r_3 is the conclusion c)
$(r_3)\, q \quad\ (p_1), (r_2)\ \text{DS}$

This example demonstrates that the conclusion follows from the premises by a small, uncomplicated number of steps, each of which was justified by reference to a basic inference rule in Table 2.13. Sometimes, however, the premises of an inference have a logical form to which no inference rule can be applied. Here is an illustration of such a situation:

$(p_1) \neg (p \vee q)$
$(p_2) \, r \Rightarrow q$
_____

$(c) \therefore \neg r$

There is no rule by which we can dissolve a negated disjunction. Similarly, there is no rule by which we can detach either part of an implication without the aid of a second proposition (either the antecedent alone or the denied consequent alone). Thus, it would appear that we cannot get started on our demonstration.

Fortunately, however, we are able to make use of a set of translations, or rules of replacement, with which we can change the original forms of our premises so that our inference rules may apply to them. Table 2.14, then, is a set of additional rules. Returning to our inference above, we can see that the first premise

$$\neg (p \vee q)$$

may be rewritten as its logically equivalent (synonymous) expression

$$\neg p \wedge \neg q$$

and the inference now proceeds validly

$(p_1) \neg (p \vee q)$
$(p_2) \, r \Rightarrow q$
_____

$(c) \therefore \neg r$
_____

$(r_1) \neg p \wedge \neg q$	(p_1) De Morgan law
$(r_2) \neg q \wedge \neg p$	(r_1) Commutativity
$(r_3) \neg q$	(r_2) Simplification
$(r_4) \neg r$	(p_2), (r_3) MT (r_4 is the conclusion c)

TABLE 2.14   RULES OF REPLACEMENT

Involution (Double Negation)	$p \Leftrightarrow \neg\,\neg\,p$
Commutativity	$(p \wedge q) \Leftrightarrow (q \wedge p)$ $(p \vee q) \Leftrightarrow (q \vee p)$
Associativity	$[p \wedge (q \wedge r)] \Leftrightarrow [(p \wedge q) \wedge r]$ $[p \vee (q \vee r)] \Leftrightarrow [(p \vee q) \vee r]$
De Morgan's Laws	$\neg\,(p \wedge q) \Leftrightarrow (\neg p \vee \neg q)$ $\neg\,(p \vee q) \Leftrightarrow (\neg p \wedge \neg q)$
Distributivity	$[p \wedge (q \vee r)] \Leftrightarrow [(p \wedge q) \vee (p \wedge r)]$ $[p \vee (q \wedge r)] \Leftrightarrow [(p \vee q) \wedge (p \vee r)]$
Equivalence	$(p \Leftrightarrow q) \Leftrightarrow [(p \Rightarrow q) \wedge (q \Rightarrow p)]$ $(p \Leftrightarrow q) \Leftrightarrow [(p \wedge q) \vee (\neg p \wedge \neg q)]$
Contraposition	$(p \Rightarrow q) \Leftrightarrow (\neg q \Rightarrow \neg p)$
Implication	$(p \Rightarrow q) \Leftrightarrow (\neg p \vee q)$
Exportation	$[p \Rightarrow (q \Rightarrow r)] \Leftrightarrow [(p \wedge q) \Rightarrow r]$
Idempotency	$(p \wedge p) \Leftrightarrow p$ $(p \vee p) \Leftrightarrow p$

The 10 rules of inference, together with the 16 rules of replacement constitute the core of the method of natural deduction. Of course, there are many other aspects of demonstrating validity, but these are beyond the scope of this book and are not required for the primary objective of our exposition of fuzzy set theory.

2.3   PREDICATE LOGIC

Propositional logic is the logic in which validity depends on the pattern of propositions as simple units of reasoning; the logical relationships between any two units of reasoning are relationships between two complete propositions. But often, we must make inferences which do not seem to depend entirely on such "external" relationships. For example, consider the following argument:

All dogs are quadrupeds
Lassie is a dog

Therefore, Lassie is a quadruped

Using our method of replacing each simple, atomic proposition with a propositional variable, we obtain this schematic representation:

$$p$$
$$q$$
$$\overline{}$$
$$\therefore r$$

But if we perform a truth-table test for validity on this inference, we will find that there exists one assignment of truth values to the propositional variables for which both premises are true and the conclusion is false. Hence, we are forced to judge this inference invalid! But the inference is obviously correct, and thus our propositional method is apparently unsuited to representing it.

The reason for this failure is that the inference about Lassie depends not on the propositions, taken as basic units, but on the terms *inside* each proposition: It is the internal structure of the premises which guarantees the validity of this inference, and in this internal structure, it is the *terms* that are basic. Further, we can see that the inference also depends on the interplay between the singular proposition 'Lassie is a dog' and the general proposition 'All dogs are quadrupeds.'

The two kinds of terms comprising the premises above are the *subject term*—for example, 'dogs'—and the *predicate term*—in the same sentence, 'are quadrupeds.' The subject term indicates the category of thing about which we are talking and the predicate term attributes a property, or predicate, to that thing. Hence, this logical representation is called *predicate logic*.

Because propositional logic cannot distinguish between subject and predicate terms, we must employ a different method to represent the internal structure of sentences, such as the ones comprising our argument above. This new method includes the structural apparatus of propositional logic, but it has enough flexibility to distinguish between *singular propositions* and *general propositions*.

Singular Propositions

The second premise, 'Lassie is a dog,' is composed of a subject term, 'Lassie,' and a predicate term, 'is a dog.' In this case, the subject term is a proper name and is represented by a lowercase letter taken from the spelling of the name. Thus. 'Lassie' is represented as 'l,' which is called an *individual constant*. The predicate term is replaced by an uppercase *predicate variable*, in this case 'D' for 'is a dog.' The result is the following:

Dl

Of course, there are many other dogs with different proper names, and they would generate these sentences and their symbolizations:

Fido is a dog   Df

Buster is a dog   Db

Ginger is a dog   Dg

and so on. One important feature of these symbolizations is that each represents a sentence which has some truth value—each is a proposition. But we also notice that all of the symbolized sentences shown above share the same structure, so we may abstract it by replacing their individual constants with *individual variables*, such as the variable "x." Accordingly, we see that these sentences are really in the form

Dx

where the meaning of the variable "x" varies with the actual individual name that could be substituted for it. "Dx" is called a *propositional function*, because it establishes a pattern for some proposition which is generated by substituting an individual constant for the variable. Since "Dx" is not itself a proposition, it does not have any truth value of its own. Using this kind of language, we can then say that the propositions 'Df,' 'Db,' and 'Dg' are true or false *substitution instances* of the propositional function "Dx."

Thus, one way to produce a proposition from a propositional function is by the substitution of individual constants for variables. This is called *instantiation*. However, it is not the only way to produce a proposition from the propositional function Dx: We may also produce propositions by so-called *generalization*.

General Propositions

Based on our knowledge of the four dogs so far named, we could say, for example, that there exists something which is a dog. This is a more general proposition than is the singular, specific proposition 'Lassie is a dog.' Further, this general proposition makes an existential claim about a thing having the property of being a dog. Hence, we say that we are forming the *existential* generalization of the original propositional function by attaching the existential quantifier, $\exists x$, to it:

$$(\exists x)\, Dx$$

which stands for 'there exists at least one x, such that the x is a dog.'

Referring back to our original inference, we can also make use of existential generalization to form a proposition from the propositional function $Dx \land Qx$:

$$(\exists x)\, (Dx \land Qx)$$

which stands for 'there exists at least one thing, such that it is both a dog and a quadruped.'

If we restrict our context—our universe of discourse—to a setting in which we refer only to a roomful of dogs, we may use the propositional function Dx as the pattern for a more sweeping generalization, namely the universally generalized claim 'everything is a dog,' by attaching the *universal quantifier*, $\forall x$, to the propositional function:

$$(\forall x)\, Dx$$

which signifies the sentence 'for any x, x is a dog' ('everything is a dog').

Further, we can make use of the universal quantifier to symbolize the first premise of our inference:

$$(\forall x)\, (Dx \Rightarrow Qx)$$

which signifies the sentence 'for any x, if x is a dog, then x is a quadruped.'

Each quantifier has jurisdiction over all its variables (all the x's) contained in the expression immediately to its right, from the opening parentheses to the corresponding closing parentheses. This jurisdiction is called the *scope of the quantifier*. In the previous expression, the scope of the quantifier $(\forall x)$ includes the variable x in both occurrences, Dx and Qx. When individual variables are within the scope of their quantifiers, they are said to be *bound*; otherwise, they are *free*. Accordingly, we can say that a propositional function is an expression containing at least one free variable. A general expression is a proposition if and only if all of its variables are bound.

Relations and Multiple Quantification

Propositions represented in the notation of predicate logic may be of great complexity and include many logical connectives. Further, they may be used to represent not only properties, but also relations. For example, the singular sentence

John loves Mary

is symbolized as

Ljm

where the predicate L is actually a relation, indicated by two consecutive individual constants. Similarly, the general sentence

Everything is attracted by something

makes use not only of a relational predicate, but also two quantifiers, each of which refers to a member of the relation:

$$(\forall x)(\exists y)Ayx$$

Quantifier Negation

The two quantifiers may be interdefined: One may be represented in terms of the other. For example, if we want to represent the sentence

It is false that everything is square

we could write

$$\neg\,(\forall x)\,Sx$$

But this negated universal proposition has the same meaning as does the existential sentence

There is something which is not square

and this is symbolized as

$$(\exists x) \neg\, Sx$$

Accordingly, we accept the so-called *quantifier negation* equivalences depicted in Table 2.15; here, the predicate variable P is used as a generic predicate.

TABLE 2.15   QUANTIFIER REPLACEMENT RULES

$(\forall x)\, Px$	$\Leftrightarrow$	$\neg\,(\exists x)\,\neg\, Px$
$\neg\,(\forall x)\, Px$	$\Leftrightarrow$	$(\exists x)\,\neg\, Px$
$(\forall x)\,\neg\, Px$	$\Leftrightarrow$	$\neg\,(\exists x)\, Px$
$\neg\,(\forall x)\,\neg\, Px$	$\Leftrightarrow$	$(\exists x)\, Px$

The Square of Opposition

One last, and important, set of inferential relationships among existential and universal propositions is an extension of the quantifier negation replacement rules above. These relationships are also studied in traditional Aristotelian logic and are often represented on the so-called *square of opposition*. This square, seen in Figure 2.2, represents the relationships among the four types of quantified propositions represented in Table 2.16.

TABLE 2.16   BASIC FORMS OF QUANTIFIED PROPOSITIONS

Universal affirmative (A)	"all S are P"	$(\forall x)\,(Sx \Rightarrow Px)$
Universal negative (E)	"no S are P"	$(\forall x)\,(Sx \Rightarrow \neg\, Px)$
Existential affirmative (I)	"there exists at least one S which is P"	$(\exists x)\,(Sx \wedge Px)$
Existential negative (O)	"there exists at least one S which is not P."	$(\exists x)\,(Sx \wedge \neg\, Px)$

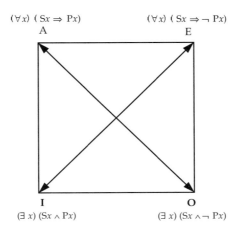

Figure 2.2   The square of opposition.

This square illustrates several truth-functional relationships among these basic classes of propositions. First, it shows that the existential negative proposition (O) is the *contradictory* of the universal affirmative proposition (A), and that the E and I propositions are also contradictories. Contradictories are propositions which are such that when one of them is true, then the other must be false. The two universal sentences, A and E are not contradictories: They may both be false and cannot both be true; they are called *contraries*. The two existential sentences are called *subcontraries*: They may both be true but cannot both be false. Further, if we assume that the things mentioned in the two universal sentences exist, then the two existential sentences may be inferred as *immediate inferences* from their corresponding universal sentences above them; for example, if the A sentence is true, then the I sentence must be true, also.

In predicate logic we accept the same set of basic inference rules and rules of replacement as in propositional logic. However, the addition of quantifiers requires a supplementary set of rules allowing us to draw quantified inferences. A discussion of these, however, is beyond the scope of this book; the reader may consult any standard introductory symbolic logic textbook for further instruction.

EXERCISES

2.1 For each of the following natural-language propositions, provide a complete symbolization, using the propositional variable p for the first atomic proposition in each exercise, q for the second, r for the third, and so on.

(a) If the exam is fair, then the class will do well.

(b) We will have a hot summer if and only if the Gulf Stream and the Alberta Clipper do not both influence the weather.

(c) It will not rain, unless a cold front passes over our valley and the humidity rises.

(d) If either the value of the dollar drops or the value of the yen rises, then although the bond market will remain flat, interests rates will increase.

(e) Both Chile and Argentina are emerging as strong economies, but neither Paraguay nor Colombia is increasing its GDP.

(f) Students will graduate from this university if and only if they have met all the distribution requirements and have maintained a "C" average.

(g) Only if the computer industry solves the problems of maintaining information privacy will intellectual property rights be secure.

(h) If neither the dependent nor the dependent's spouse is required to file BUT they file a joint return, then you may claim him (or her), if the other four tests are met.

(i) Germany and Italy will not both win the soccer World Cup.

(j) It is not true that if you wear a copper bracelet, then your arthritis will disappear.

2.2 Given that p, r, and t are true (T) and that q, s, and v are false (F), determine the truth values of each of the following compound propositions:

(a) $(p \Rightarrow q) \Rightarrow (\neg p \Rightarrow \neg q)$

(b) $[(p \wedge q) \Rightarrow s] \Rightarrow [p \Rightarrow (q \Rightarrow s)$

(c) $[r \wedge (s \vee t)] \Rightarrow [(r \wedge s) \vee (r \wedge t)]$

(d) $\neg\{[(p \Leftrightarrow q) \wedge (r \Leftrightarrow s)] \wedge (p \vee r)\} \Rightarrow (q \vee s)$

(e) $(p \Rightarrow v) \Rightarrow \{[p \Rightarrow (v \Rightarrow q)] \Rightarrow (p \Rightarrow q)$

2.3 Using full truth tables, determine the truth value of each of the propositions in Exercise 2.2 under all possible assignments of truth values to their component atomic propositions. Study these truth tables and consider, whether or not they reveal any interesting information about the complex propositions in Exercise 2.2 that was not obvious, when you were using the truth values given in that exercise.

2.4 Using a truth table, determine whether each of the following propositions is a tautology, a contradiction, or a contingent proposition.

(a) $[(p \wedge q) \Rightarrow p] \Rightarrow q$

    (b) $(p \Rightarrow q) \Rightarrow (q \Rightarrow p)$
    (c) $\neg\{(p \Leftrightarrow q) \Leftrightarrow [(p \wedge q) \vee (\neg p \wedge \neg q)]\}$
    (d) $[(p \Leftrightarrow \neg q) \wedge s] \wedge (\neg p \Leftrightarrow q)$
    (e) $\{[p \Rightarrow (r \vee s)] \wedge [q \Rightarrow (r \Rightarrow q)]\} \Rightarrow (p \vee q)$

2.5 Using a truth table, determine whether any of the following expressions is a logical equivalence:
    (a) $(p \Rightarrow q) \Leftrightarrow (\neg q \Rightarrow \neg p)$
    (b) $\neg(p \wedge q) \Leftrightarrow (\neg p \vee \neg q)$
    (c) $[p \wedge (q \vee r)] \Leftrightarrow \neg[(p \wedge q) \vee (p \wedge r)]$
    (d) $(p \Rightarrow q) \Leftrightarrow (q \Rightarrow p)$

2.6 For each of the following inferences, identify the basic inference rule of which it is a substitution instance:
    (a) $(p \wedge \neg q) \Rightarrow (r \vee s)$

        $\dfrac{p \wedge \neg q}{\therefore r \vee s}$

    (b) $(\neg p \Rightarrow q) \wedge (r \vee \neg s)$

        $\overline{\therefore [(\neg p \Rightarrow q) \wedge (r \vee \neg s)] \vee (\neg p \Rightarrow s)}$

    (c) $[(p \Leftrightarrow \neg q) \wedge s] \wedge (\neg p \Leftrightarrow q)$

        $\overline{\therefore (p \equiv \neg q) \wedge s}$

    (d) $r \Rightarrow (s \Leftrightarrow \neg p)$
        $(s \Leftrightarrow \neg p) \Rightarrow q$

        $\overline{\therefore r \rightarrow q}$

    (e) $(\neg r \Leftrightarrow s) \vee (t \vee u)$
        $\neg(\neg r \Leftrightarrow s)$

        $\overline{\therefore t \vee u}$

2.7 Demonstrate the validity of each of the inferences in Exercise 2.6 by using a truth table to identify the statement form of the conditional proposition that describes each inference

2.8 Symbolize each of the following natural-language propositions using the suggested notation and making the symbolization reflect the sense of each English proposition as much as possible.
    (a) All bald eagles are on the Endangered Species list and are protected. (Bx: x is a bald eagle; Ex: x is on the Endangered Species list; Px: x is protected.)
    (b) Some mutual funds are appropriate for risk-averse investors. (Mx: x is a mutual fund; Rx: x is appropriate for risk-averse investors.)
    (c) Not all politicians campaign using negative ads, but a few politicians engage in negative campaign tactics. (Px: x is a politician; Nx: x campaigns using negative ads; Tx: x engages in negative campaign tactics.)
    (d) Some students are industrious only if their parents buy them a new car. (Sx: x is a student; Lx: x is industrious; Rx: x receives a new car from his/her parents.)

(e) If John plays baseball, then there exists at least one person who plays a sport. (B*x*: *x* plays baseball; P*x*: *x* is a person; S*x*: *x* plays a sport.)

(f) Some medical schools are popular and expensive only if they are famous and are located in a warm climate. (M*x*: *x* is a medical school; P*x*: *x* is popular; E*x*: *x* is expensive; F*x*: *x* is famous; W*x*: *x* is located in a warm climate.)

(g) Either almost all traditional accounting methods are maintained in the curriculum, or a few new computers are required and some new software packages are recommended. (T*x*: *x* is a traditional accounting method; M*x*: *x* is maintained in the curriculum; C*x*: *x* is a computer; Q*x*: *x* is required; S*x*: *x* is a software package; R*x*: *x* is recommended.)

(h) If only roses are fragrant, then either no hollyhocks are fragrant or some flowers are not roses. (R*x*: *x* is a rose; F*x*: *x* is fragrant; L*x*: *x* is a flower; H*x*: *x* is a hollyhock.)

(i) If anyone values her education, then so does everyone. (P*x*: *x* is a person; V*x*: *x* values her (his) education.)

(j) Most comets are harmless if and only if their orbits do not lie between the Earth and the Moon and they are not composed primarily of iron. (C*x*: *x* is a comet; H*x*: *x* is harmless; O*x*: *x* is an orbit of a comet; L*x*: *x* lies between the Earth and the Moon; I*x*: *x* is composed primarily of iron.)

3

CLASSICAL SET THEORY

One of the most important aspects of classical propositional logic is that it is designed to help us make distinctions. Accordingly, we distinguish propositions according to their logical form; we distinguish tautologies, contradictions, and contingent propositions from one another, and—very importantly—we crisply distinguish truth from falsity. Predicate logic extends our ability to distinguish: It allows us to talk about individuals or groups of them in terms of the properties, or predicates, they possess. Specifically, quantifiers help us to distinguish whether a given property is satisfied by all individuals in a designated group or just some of them.

Classical set theory is another form of representing the same kind of distinctions; namely, the distinctions we make between and among groups of things that we perceive to share a characteristic or property. As in classical propositional and predicate logic, classical set theory is founded on the idea that we can make crisp, exact distinctions between two groups; according to it, we should always be able to tell exactly whether an individual is definitely in the group or definitely outside the group. Of course, in most situations, we are not able to circumscribe an individual's set membership precisely, but in dealing with the basic components of set theory, we must accept its assumption that groups—or sets—have sharp boundaries.

3.1 BASIC CONCEPTS AND NOTATION

The concept of a set is one of the most basic concepts in both logic and mathematics. By saying that it is basic, we mean there are no more basic concepts in terms of which a "set" may be defined. However, we can give a nontechnical description of what counts as a set and its properties. We use the term 'set' to refer to any collection of items or individuals. Collections can be composed of any sort of thing: buildings, students, cars, hamburgers, odd numbers, letters of the alphabet, and so on.

We can say that a set is a collection of things that can be distinguished from one another as individuals and that share some property. Each individual in this collection is called a *member*, or *element*, of the *set*. Throughout this book, we use the uppercase letters A, B, C,..., X, Y, Z to denote sets. Accordingly, uppercase letters will function as the labels for sets. Further, we use the lowercase letters a, b, c,..., x, y, z to denote elements of sets.

If the individual a belongs to the set A, we write this "belonging" relationship using the notation $a \in A$. The symbol $\in$ is read "is an element of." If an element a is not a member of the set A, we express this fact with the symbol $\notin$. In classical set theory, there are only two possible relationships between the individual a and the set A: either $a \in A$, or $a \notin A$. That is, a is either an element of the set A, or it is not an element of the set A; it is either definitely "in" or it is definitely "out." This is illustrated in Fig. 3.1.

In addition, we use the standard symbols $\exists$ and $\forall$, respectively, for the existential and the universal quantifiers. The expression $\forall x \in A$ represents the beginning of a universal proposition: "for any element x in set A." Similarly the expression $\exists x \in A$ represents the beginning of an existential proposition: "there exists at least one element x in set A."

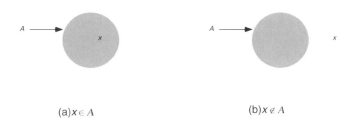

(a)$x \in A$ (b)$x \notin A$

Figure 3.1   In the set and not in the set.

There are several methods used to represent or describe sets, the most common being the *list method* and the *rule method*. The list method is usually used for small sets. Using it, we represent a set by *enumerating* its elements, enclosing them with a pair of braces. For example, the set A whose members are a, b, and c is represented in this way:

$$A = \{a, b, c\}$$

Other examples are

$B = \{$John, Barbara, George, Nana$\}$
$C = \{a, b, c, ..., x, y, z\}$
$D = \{$penny, nickel, dime, quarter, half-dollar, dollar$\}$

For each of these sets, the items, whose names are cited within each pair of braces, constitute a set bearing the label, or name, indicated to the left of the equality sign. One characteristic of these sets is that we can fairly easily determine how many members each has: Set B has 4 members, set C has 26, and set D has 6. The number of members of a finite set is its *size*, which is usually referred to as the *cardinality* of the set. Hence, set B, for example, has a cardinality of 4, and this property is written using vertical bars around the set name: $|A| = 4$. One feature of sets is that it is possible for a set to have only one member; a one-member set is called a *singleton*. For example, the set of all currently serving United States presidents is a set containing only one member. If we label that individual as p, then this set will be written as

$$\text{President} = \{p\}$$

Using the *rule method*, a set, say C, can be represented in a way that stipulates a rule whereby we can form the desired set:

$$C = \{x \mid P(x)\}$$

This expression is read, "the set C is composed of elements x, such that (every) x has the property P" (the symbol $\mid$ is read as "such that"). This representation says that the set C is constituted by elements—indifferently labeled x—which all share the property or properties mentioned within the braces. It is required in the classical set theory that the property P be such that for any given individual x, the proposition $P(x)$ be either true or false.

The following are examples of sets defined by the rule method:

$C = \{x \mid x$ is a legal United States coin type in 1994$\}$
$E = \{x \mid x$ is a real number between 10 and 100$\}$
$F = \{x \mid x$ is an integer$\}$

A set whose elements are themselves sets, which is usually referred to as a *family of sets*, can be defined in the form

$$\{A_i \mid i \in I\}$$

where i and I are called the *index* and the *index set*, respectively. In this text, families of sets are denoted by capital script letters, such as $\mathcal{A}, \mathcal{B}, \mathcal{C}$, and so on.

Universal Set and Empty Set

In the context of each application of fuzzy set theory, we recognize one particular set as a *universal set*. This is a set that consists of all the individuals that are of interest in that application. For example, if we are interested in classifying students on our campus by various criteria, then the universal set consists of all students on our campus; in plane analytic geometry, the universal set is the set of all ordered pairs of real numbers; for certain linguistic studies, the universal set may be the set of all letters in the alphabet and all punctuation signs, and so on. In this book, we usually use the letter X to denote the universal set.

Just as we have a set containing all the individuals of interest in our universal set, so we also have a set containing nothing at all: the aptly named *empty set*. The empty set is a set which contains no elements and is denoted by the symbol $\varnothing$. It is something like a "shell" having a description that does not happen to fit anything. For example, the set "living unicorn" is an empty set.

Set Inclusion

Everyday contexts teach us that many sets are themselves parts of larger sets. For example, the set of college freshmen on our campus is part of the larger set of undergraduate students on the campus. This means that every member of the set of freshmen is also a member of the set of undergraduate students. Now, assume that A and B are sets. If every member of set A is also a member of set B, then A is called a *subset* of B. In other words, being a member in A implies being a member in B (if the members are x's, then this definition says that $x \in A$ implies that $x \in B$). We represent the subset relation as

$$A \subseteq B$$

One important feature of this relation is that *every set is a subset of itself.*

A useful graphic representation of sets, known as a Venn diagram, is often used for visualizing relations existing among various subsets of a given universal set. In a Venn diagram, the universal set X is represented by the points within a rectangle and subsets of X are represented by circles or other simple regions inside the rectangle. For example, the relation among the set of all students on our campus, which is the universal set X in this case, and two of its subsets are shown in Fig. 3.2.

Building on the concept of subsethood, we can see that if $A \subseteq B$ and $B \subseteq A$, then A and B contain all the same elements. In this case, A and B are called *equal sets* and their equality relation is represented by

$$A = B$$

To indicate that A and B are not equal, we write

$$A \neq B$$

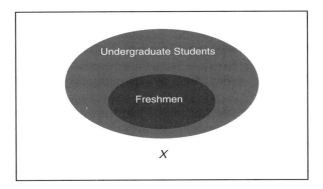

Figure 3.2   Subsets of universal set X.

If both $A \subseteq B$ and $A \neq B$, then B contains at least one individual that is not an element of A. In this case, A is called a *proper subset* of B, and this relation is represented by the expression

$$A \subset B$$

Given a universal set X, the subset relationship

$$\varnothing \subseteq A \subseteq X$$

where A is any set defined in terms of elements of X, is clear.

Power Set

The set which consists of all possible subsets of a given set X is called a *power set* of X and is denoted by the symbol

$$\mathcal{P}(X)$$

This symbol is thus an abbreviation for the set

$$\{A \mid A \subseteq X\}$$

Clearly, the propositions $A \in \mathcal{P}(X)$ and $A \subseteq X$ are equivalent.

When X is a finite set containing n elements (i.e., $|X| = n$), the number of subsets of X is 2^n; that is,

$$|\mathcal{P}(X)| = 2^n$$

This number can be easily derived by realizing that there are two possibilities for each element of X with respect to the various subsets of X: The element is either included in a given subset or it is not included in it. Hence, for all n elements, we have

$$\underbrace{2 \times 2 \times \ldots \times 2}_{n - \text{times}} = 2^n$$

possibilities. For example, when $X = \{a, b, c\}$, we have

$$\mathcal{P}(X) = \{\varnothing, \{a\}, \{b\}, \{c\}, \{a, b\}, \{a, c\}, \{b, c\}, X\}$$

3.2   *SET OPERATIONS*

In everyday life, we know that sets of things often are related in certain ways. For example, the set of all male graduate students is mutually exclusive from the set of all female graduate students. However, taken together, the composite set formed from these two is the set of all graduate students of either gender. Another way to look at these two sets is to recognize that if we take the university community to be the *universal set*, then we could divide it into the set of all graduate students and the remaining set made up of everybody who is not a graduate student. In this latter set, we would find all of the undergraduate students, all the nonmatriculated students, the faculty, the administrators, the support staff, and so on.

Each of these relationships may be abstracted from specific examples, such as the one just described, and each relationship may be more precisely defined. To study each one, let us consider a universal set X and some of its subsets, A, B, and C. In our example, the universal set X would be the set of all members of the university community; A might be the set of all graduate students, B the set of all faculty members, and C the set of all administrators. Of course, A could be divided into two of its own subsets, namely the set of all male graduate students and the set of all female graduate students, and so forth. The relationship among the various subsets of X can be studied in terms of some suitable operations defined on sets. In this section, we introduce four operations on sets: set complement, union, intersection, and difference.

Complement

The *complement*, or *absolute complement*, of a given set A, denoted by the expression $\overline{A}$, is the set of all elements in the universal set X which are *not* in A. Placing the bar over the designator for the set, A in this case, indicates that we have performed the operation of complement on the set A. For example, the complement of the set of all graduate students would be the set composed of everyone in the university community who is not a graduate student, and it would contain all the undergraduates, the faculty, the administrators, and the staff. More precisely,

$$\overline{A} = \{x \mid x \in X \text{ and } x \notin A\}$$

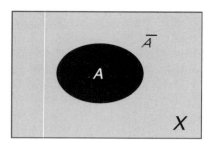

Figure 3.3   Set A and its complement.

Because a set and its complement divide the universal set exhaustively into two parts, we can see that the *complement of the complement of A* is just A itself, that is, $\bar{\bar{A}} = A$. This property of the complement, called *involution,* is analogous to the property of *double negation* in symbolic logic, since the negation of an already negated proposition is the proposition itself. Further, the complement of the empty set is the universal set: $\bar{\varnothing} = X$. Conversely, the complement of the universal set is the empty set: $\bar{X} = \varnothing$.

The concept of the set complement is illustrated by the Venn diagram in Fig. 3.3, where the dark area represents set A and the gray area represents the complement, $\bar{A}$, of A.

Union

The *union* of set A and set B—represented by the expression $A \cup B$—is the set containing all the elements belonging either to A or to B, or to both. This operation is analogous to the logical operation of inclusive disjunction. In our example, the union of the set of faculty members and the set of administrators contains all the faculty members, all the administrators, and also anyone who is both a faculty member and an administrator. Formally,

$$A \cup B = \{x \mid x \in A \text{ or } x \in B\}$$

A Venn representation is depicted in Fig. 3.4, where the circles represent sets A and B, and the grey area represents $A \cup B$.

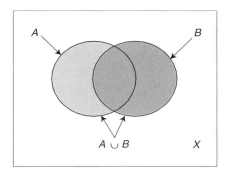

Figure 3.4   Union of sets.

The union of any set A with the universal set X just yields X; that is, $A \cup X = X$. More generally, if $B \subseteq A$, then $A \cup B = A$. In particular, $A \cup \varnothing = A$ for any set A. The union of any set A with its complement $\overline{A}$ yields the universal set X, because all the elements of the universal set must belong either to a set A or its complement $\overline{A}$. We remember this principle from classical logic, where it is known as the *law of excluded middle*. This law is expressed in set-theoretic terms as

$$A \cup \overline{A} = X$$

Intersection

The *intersection* of sets A and B, denoted by $A{\cap}B$, is the set containing all the elements belonging to both sets A and B simultaneously. It is analogous to the operation of conjunction in classical logic. Returning to our university example, we might consider the set of all female students A and the set of all graduate students B. Their intersection $A \cap B$ contains individuals who are both female students and graduate students. Thus, in the intersection of two sets, the elements have the properties of *both* sets. Formally,

$$A \cap B = \{x \mid x \in A \text{ and } x \in B\}$$

A Venn diagram representation is shown in Fig. 3.5, where the circles represent sets A and B, and the black area represents $A \cap B$.

The intersection of any set A with the universal set X yields the original set A itself, since the section in which the universal set X overlaps the set A is just the section which lies within the boundaries

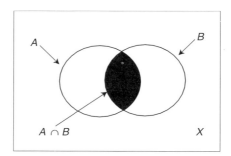

Figure 3.5   Intersection of sets.

of A. More generally, if $A \subseteq B$, then $A \cap B = A$. In particular, $\varnothing \cap A = \varnothing$ for any set A. Moreover, the intersection of any set A and its complement $\bar{A}$ is the empty set since A and $\bar{A}$ do not overlap. That is,

$$A \cap \bar{A} = \varnothing$$

which is a set-theoretic counterpart of the *law of contradiction* in propositional logic.

An additional concept involving the operation of intersection is the concept of *disjoint sets*. Any two sets A and B are said to be *disjoint*, if they have no elements in common. For example, the set of dogs and the set of cats are disjoint sets, since they have no common members. We then say that if A and B are disjoint, then their intersection is empty: $A \cap B = \varnothing$. Any set and its complement are an example of disjoint sets.

Difference

The *difference* of sets A and B is a set that consists of all the elements which belong to A, but which do not belong to B. The difference set is represented by the expression $A - B$. Formally,

$$A - B = \{x \mid x \in A \text{ and } x \notin B\}$$

Note that the difference operation is not symmetrical: $A - B \neq B - A$. Figure 3.6 illustrates the difference $A - B$, while Fig. 3.7 illustrates the difference $B - A$.

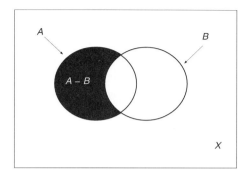

Figure 3.6   Difference of set A and set B.

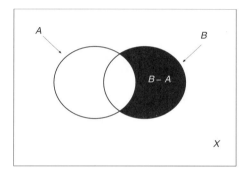

Figure 3.7   Difference of set B and set A.

The special difference of the universal set X and its subset A is $\overline{A}$, or, $X - A = \overline{A}$. However, $A - X = \varnothing$. More generally, if $A \subseteq B$ then $A - B = \varnothing$.

3.3   FUNDAMENTAL PROPERTIES

When two or more sets are combined by the use of union, intersection, and complement, these complex combinations take on some important and revealing properties. These are properties that we also encountered in propositional logic when we combined simple propositions into complex propositions by the use of the five logic operators. When combinations of sets through complements, intersections, and unions acquire new properties, we sometimes say that these properties are *satisfied* by these operations.

Our first important, and intuitive, property of sets is *involution*, whose counterpart in classical logic is the rule of double negation: $\neg\neg P \Leftrightarrow P$. Just as this rule asserts that the double negation of a proposition is equivalent to the proposition itself, so the rule of involution says that the complement of the complement of a set is just the set itself:

$$\overline{\overline{A}} = A$$

Another fundamental property of classical sets is that a set and its complement do not overlap. This means that $A \cap \overline{A} = \varnothing$ for any set acceptable by classical set theory. Observe that this property is a set-theoretic counterpart of the law of contradiction in classical logic. Moreover, each element of the universal set is either in set A or in its complement $\overline{A}$; that is, $A \cup \overline{A} = X$, which is a set-theoretic counterpart of the law of excluded middle in classical logic. By and large, these properties do not hold for fuzzy sets.

Commutativity, Associativity, and Idempotence

When two sets are combined by either the operation of union or the operation of intersection—$A \cup B$ or $A \cap B$—it does not matter in which order the two operands occur. We say that union and intersection are *commutative*, just as conjunction and disjunction are. Accordingly:

$$A \cup B = B \cup A \text{ and } A \cap B = B \cap A$$

Union and intersection are also *associative*. What this means is that when more than two sets are combined either with only union operators or only intersection operators, the placement of parentheses, grouping any two sets together, has no effect on the overall value of the combined set. Thus the order in which the union or intersection operations are performed does not matter. For example, for three sets $A, B,$ and $C,$

$$(A \cup B) \cup C = A \cup (B \cup C)$$

and

$$(A \cap B) \cap C = A \cap (B \cap C)$$

Because of the associativity of the union operation and the intersection operation, the notation for unions and intersections of several sets can be simplified. We do not need parentheses to indicate the order of operations to be performed. For example, the union of three sets A, B, and C can be denoted by $A \cup B \cup C$ and, the intersection of the three sets can be represented by $A \cap B \cap C$. More generally, the union of all sets in the family $\{A_1, A_2,..., A_n\}$ can be denoted by $A_1 \cup A_2 \cup ... \cup A_n$ or, in a simplified form, by

$$\bigcup_{i=1}^{n} A_i$$

Similarly, the intersection of the sets can be denoted by $A_1 \cap A_2 \cap ... \cap A_n$ or

$$\bigcap_{i=1}^{n} A_i$$

Note that *associativity* is a property only of expressions which contain one kind of operation, not of expressions with mixed operations (both union and intersection).

Idempotency also holds for both union and intersection. According to this property, the union of any set with itself, as well as the intersection of any set with itself yields the original set. This is a useful property, because it allows us to "collapse" redundant strings of unions and intersections, just as we are able to collapse redundant propositions that are conjoined or disjoined. Accordingly,

$$A \cup A = A$$
$$A \cap A = A$$

Distributivity

The so-called *law of distribution*—which is also an important rule in logic—allows us to "distribute" a set on one side of a union operator over the intersection of two other sets (conversely, we may also distribute a set on one side of an intersection operator over the union of two other sets). The result will show that the original main operator will become its opposite, and the original subsidiary operator will similarly become its opposite:

$$A \cap (B \cup C) = (A \cap B) \cup (A \cap C)$$
$$A \cup (B \cap C) = (A \cup B) \cap (A \cup C)$$

De Morgan's Laws

De Morgan's laws—named after the nineteenth-century British mathematician Augustus De Morgan—are rules also found in classical logic. They are useful, because, if we are dealing with the complements of sets, these laws allow us to transform an intersection of two sets into their union, and vice versa. De Morgan's laws state that the complement of the intersection of two sets is equivalent to the union of their individual complements. Similarly, the complement of the union of two sets is equivalent to the intersection of their individual complements:

$$\overline{A \cap B} = \bar{A} \cup \bar{B}$$

$$\overline{A \cup B} = \bar{A} \cap \bar{B}$$

An interesting and useful result is obtained, when we consider using both the rule of involution and De Morgan's laws at the same time:

$$A \cap B = \overline{\bar{A} \cup \bar{B}}$$

$$A \cup B = \overline{\bar{A} \cap \bar{B}}$$

The introduced properties of the operations of set complement, union, and intersection are summarized in Table 3.1. Notice that these properties exhibit the general principle of *duality*: To each property, there corresponds a dual property—something like a mirror image—obtained by replacing $\varnothing$, $\cup$, $\cap$ with X, $\cap$, $\cup$, respectively. Moreover, most of the properties in Table 3.1 are the set-theoretic counterparts of the properties that function as rules of replacement in symbolic logic, which are shown in Table 2.14.

TABLE 3.1 BASIC PROPERTIES OF CLASSICAL SET OPERATIONS

Involution	$\overline{\overline{A}} = A$
Commutativity	$A \cap B = B \cap A, A \cup B = B \cup A$
Associativity	$A \cap (B \cap C) = (A \cap B) \cap C, A \cup (B \cup C) = (A \cup B) \cup C$
Distributivity	$A \cap (B \cup C) = (A \cap B) \cup (A \cap C),$ $A \cup (B \cap C) = (A \cup B) \cap (A \cup C)$
Idempotence	$A \cap A = A, A \cup A = A$
Absorption	$A \cap (A \cup B) = A, A \cup (A \cap B) = A,$
Absorption by $\varnothing$ and X	$A \cup X = X, A \cap \varnothing = \varnothing$
Identity	$A \cap X = A, A \cup \varnothing = A$
Law of contradiction	$A \cap \overline{A} = \varnothing$
Law of excluded middle	$A \cup \overline{A} = X$
De Morgan laws	$\overline{A \cap B} = \overline{A} \cup \overline{B}, \overline{A \cup B} = \overline{A} \cap \overline{B}$

3.4 *CHARACTERISTIC FUNCTIONS OF CRISP SETS*

Two methods for representing sets were introduced in Sec. 3.1: the list method and the rule method. Another method for representing sets is based on the so-called characteristic functions. This method is particularly important in this text since, contrary to the other two methods, it can be generalized to fuzzy sets. Before describing the method, let us introduce first the mathematical concept of a function.

A *function* is an assignment of elements of one set to elements of another set. What this means is that each element of the first set is paired with an element of the second. The elements of the second set can be said to be the images (or values) of elements in the first. For example, if we have a symbolic complex proposition containing three propositional variables, then they constitute a set having three members. The set of two possible truth values is a second set. When we stipulate the truth values of each of the three propositions, we assign

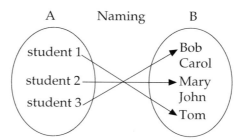

Figure 3.8   Example of a function called *Naming*.

to each proposition—each member of the first set—an element in the second set—one of the two truth values.

Again, a function *f* assigns to each element in a given set, say set *A*, an element in another set, say set *B*. This assignment has to satisfy two criteria. One is that every element in *A* must be assigned an element in *B*. The other is that each element in *A* can be assigned only one element in *B*. If an assignment fails to pair an element in *A* with an element in *B*, then this assignment is not a function. Similarly, if an element in *A* is matched with more than one element in *B*, it is also not a function.

Consider all the students in a classroom to be a set *A*. We will let set *B* include all the names of the students in *A* and also the names of other students. Then, the correspondence between students in *A* and names in *B* is a function. Suppose there are three students in the classroom set *A*, say, student 1, student 2, and student 3. The set of names *B* {John, Tom, Bob, Mary, Carol} includes the names of the students in *A*. The correspondence between each of the three students and their three names in set *B*, respectively, is called *naming* and is a function.

Assume that student 1 is Tom, that student 2 is Mary, and that student 3 is Bob. Then, Fig. 3.8 shows an example of *naming*.

A function *f* from a set *A* to a set *B* is usually denoted by $f: A \to B$. Note that for an assignment to be a genuine function, it is not necessary that each element in *B* be assigned to an element in *A*. For instance, the name Carol is not used as an assignment in the naming function. If it happens that all elements in *B* are assigned to elements in *A*, then this function *f* is called an *onto* function. Thus, our naming function is not *onto*. It is also not necessary that all elements in *A* be assigned different elements in *B*. If two or more elements in *A* have the same element in B as their assigned value, then the function determining this kind of assignment is called *many-to-one*. On the other hand, when elements in B are paired with no more than one element in A, we have a *one-to-one* function. Accordingly, the naming function discussed here is one-to-one (note that in your class you may not obtain a one-to-one naming function because there may be two students with the same name). The concept of a function plays a very important role in mathematics. Here, we just introduce the concept in a simplified way for our purposes.

Assume now that X is a universal set. Then, every subset of X can be uniquely represented by a function from X to the set {0, 1}. This unique function is called the *characteristic function of the subset*. It is based on a general property shared by all subsets: An element of the universal set X is either in a given subset or not: being IN or OUT is the only feature of an element that is important to the definition of the subset. Let A be any subset of X. Then, its characteristic function, denoted by χ_A, is defined for each $x \in X$ by the following rule:

$$\chi_A(x) = \begin{cases} 1 \text{ if } x \in A \\ 0 \text{ if } x \notin A \end{cases} \tag{3.1}$$

for any $x \in X$.

For example, suppose X is the set of all nonnegative real numbers and A is the set of real numbers from 5 to 10. Then, A is a subset of X whose characteristic function is defined for each x by the rule

$$\chi_A(x) = \begin{cases} 1 \text{ if } 5 \leq x \leq 10 \\ 0 \text{ otherwise} \end{cases}$$

This function is illustrated graphically in Fig. 3.9, where $\circ$ and $\bullet$ denote that the value is excluded or included, respectively.

It is important to realize that the values 1 and 0 involved in characteristic functions have no numerical significance. They are only convenient symbols by which elements of a designated universal set X that belong to a particular subset of X are distinguished from those not belonging to it.

When representing sets by their characteristic functions, some set-theoretic concepts can be described functionally. For example, a set A is a subset of another set B if and only if the characteristic function

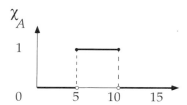

Figure 3.9   Characteristic function of
the set of real numbers from 5 to 10.

χ_A is less than or equal to the characteristic function χ_B. Here, a function is less than or equal to another function if the value of the first function is less than or equal to the value of the second function at each element in the universal set. Mathematically, we state this definition in the following way:

$$A \subseteq B \text{ if and only if } \chi_A(x) \leq \chi_B(x) \text{ for each } x \in X \qquad (3.2)$$

Another reason to introduce characteristic functions is that the set operations discussed in Section 3.2 can be represented by operations on characteristic functions. For example, the characteristic function $\chi_{\bar{A}}$ of the complement of a set A, can be obtained for each $x \in X$ from the characteristic function χ_A, according to the equation

$$\chi_{\bar{A}}(x) = 1 - \chi_A(x) \qquad (3.3)$$

The characteristic functions of the union and intersection of two sets, A and B, can also be obtained from the characteristic functions of these sets by applying the formulas

$$\chi_{A \cup B}(x) = \max(\chi_A(x), \chi_B(x)) \qquad (3.4)$$

$$\chi_{A \cap B}(x) = \min(\chi_A(x), \chi_B(x)) \qquad (3.5)$$

for each $x \in X$.

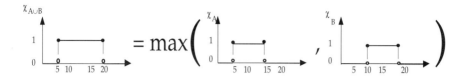

Figure 3.10   Characteristic function of the union of two sets.

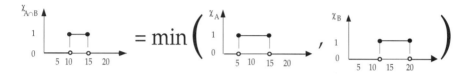

Figure 3.11   Characteristic function of the intersection of two sets.

For example, let A be the set of real numbers from 5 to 15 and let B be the set of real numbers from 10 to 20. Then, the union of A and B is the set of all real numbers from 5 to 20. The graphical illustration of Eq. (3.4) for this example is shown in Fig. 3.10. Similarly, the intersection of A and B is the set of all real numbers from 10 to 15. The graphical illustration of Eq. (3.5) in this example is shown in Figure 3.11.

3.5   *OTHER CONCEPTS*

In this section, we introduce some additional mathematical concepts that are needed in this book. One of these is the concept of a *real number*, informally encountered in the previous section. The set of real numbers plays a very important role in this book. Indeed, this concept is used almost everywhere in the discussion of fuzzy sets. The set of all real numbers is usually denoted by $\mathbb{R}$. It is a set with a so-called *total ordering* or *linear ordering*; that is, any two real numbers can be ordered according to the usual sense of the expression "less than or

Figure 3.12   The axis of real numbers.

equal to." The real number set $\mathbb{R}$ is often represented by an x-axis, with each point on the axis corresponding to a real number in $\mathbb{R}$. Fig. 3.12 illustrates this x-axis. This geometric representation of real numbers is usually called the *real line* or the *real axis*.

The set of all points between given points a and b on the real line $(a \leq b)$ is called an *interval*. An interval that includes the endpoints, a and b, is called a *closed interval* and is denoted by $[a, b]$. An interval that does not contain the endpoints is called an *open interval* and is denoted by (a, b). Intervals that include only one of the endpoints are called *half-open intervals*; the one that includes only a is denoted by $[a, b)$, and the one that includes only b is denoted by $(a, b]$.

For any two real numbers on the x-axis, the one on the left is less than the one on the right. This x-axis is also called the *one-dimensional Euclidean space*. The Cartesian product of two real number sets, denoted by $\mathbb{R} \times \mathbb{R}$, is the *two-dimensional Euclidean space*, usually called a *plane*. The two-dimensional Euclidean space can be represented by so-called *Cartesian coordinates*, or x-y axes, shown in Fig. 3.13.

More generally, the Cartesian product of n real number sets $(n \geq 1)$ is called the *n-dimensional Euclidean space*. In this book, however, we deal only with Euclidean spaces whose dimensions are not greater than 2.

The *Cartesian product*—named after the philosopher-mathematician René Descartes—of any arbitrary sets, say A and B (in this order), is the set of all possible ordered pairs constructed in such a way that the first element in each pair is a member of A and the second element is a member of B.

An ordered pair composed from elements a and b—represented as $\langle a, b \rangle$—is a simple array of two individuals arranged in such a way that it matters which one appears as the first member of the pair. That is, $\langle a, b \rangle$ is a different pair from $\langle b, a \rangle$. Elements in an ordered pair may be equal; this happens when sets A and B share some elements and, in particular, when $A = B$.

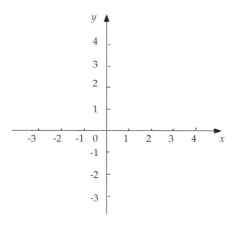

Figure 3.13   Two-dimensional coordinate system.

The Cartesian product of sets A and B, denoted by $A \times B$, is formally defined by the equation

$$A \times B = \{\langle a, b \rangle \mid a \in A \text{ and } b \in B\}$$

Imagine two sets of individuals divided according to whether they are males (A) or females (B) and participating in a dance, where they all have to pair up. We wish to construct a Cartesian product whose pairs are all possible matchings of prospective dance partners. They will be ordered, in that the first member of each pair will be male. Let

$$A = \{\text{John, Bill, Sam}\}$$
$$B = \{\text{Lisa, Mary, Donna}\}$$

Then,

$$A \times B = \left\{ \begin{array}{l} \langle\text{John, Lisa}\rangle,\langle\text{John, Mary}\rangle,\langle\text{John, Donna}\rangle,\langle\text{Bill, Lisa}\rangle,\langle\text{Bill, Mary}\rangle \\ \langle\text{Bill, Donna}\rangle,\langle\text{Sam, Lisa}\rangle,\langle\text{Sam, Mary}\rangle,\langle\text{Sam, Donna}\rangle \end{array} \right\}$$

An important property of some subsets of a Euclidean space is called *convexity*. A subset of a Euclidean space A is *convex*, if line segments between all pairs of points in the set A are included in the set. Formally, A is *convex* if and only if for any $\mathbf{r}, \mathbf{s} \in A$, and any $\lambda \in [0,1]$,

$$\lambda\mathbf{r}+(1 - \lambda)\mathbf{s} \in A$$

In one-dimensional Euclidean space, points r and s are real numbers. In two-dimensional Euclidean space, they are ordered pairs; and, in n–dimensional Euclidean space, they are ordered n–tuples, representing values of n coordinates. We use bold letters $\mathbf{r}$ and $\mathbf{s}$ to indicate that they represent, in general, ordered tuples. This notation is common in the literature.

Different values of λ in the closed interval $[0,1]$ are used in this definition to distinguish points on the straight line segment connecting $\mathbf{r}$ and $\mathbf{s}$. All points on the segment are covered by using all values (real numbers) for λ in $[0,1]$. For each value of $\lambda \in [0,1]$, the term $\lambda\mathbf{r}+(1 - \lambda)\mathbf{s}$ stands for a particular point lying on the segment.

For one-dimensional Euclidean space points r and s are real numbers. Assume that $r \leq s$. When $\lambda = 0$, then $\lambda r+(1 - \lambda)s = s$, and we obtain the right endpoint of the segment connecting r and s. When $\lambda =1$, then $\lambda r+(1 - \lambda)s = r$, and we obtain the left endpoint of the segment. When λ is between 0 and 1, we obtain a point between r and s. For example, when $\lambda = 0.5$, we obtain $0.5r + 0.5 s = (r + s)/2$; this is the point in the center of the segment. The meaning of the expression $\lambda r+(1 - \lambda)s$ is illustrated in Fig. 3.14 for $r = -2$, $s = 3$, and $\lambda = 0$, 0.3, 0.5, 0.8, 1.

Clearly a convex subset of the real line must be an *interval* of real numbers. For example, $[0, 1]$, $(-10, 20)$, $(-3, 0]$ and $[2, 3)$ are convex sets. A set of real numbers which is formed by two separated intervals is not convex. For example, the set $[0,1] \cup [2,3]$ is not convex, because there exist two numbers, r and s, in this set such that the number $\lambda r + (1 - \lambda) s$ for a suitable value of $\lambda \in [0,1]$ is not in the set.

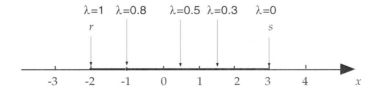

Figure 3.14   The meaning of the expression $\lambda r+(1-\lambda)s$ in one-dimensional Euclidean space.

For example, let $r = 1$, $s = 2$, and $\lambda = 0.5$. Then, $\lambda r + (1 - \lambda) s = 1.5$ and, clearly, the number 1.5 is not included in the given set $[0,1] \cup [2,3]$.

    For two-dimensional Euclidean space, each point is characterized by a pair of real numbers—the x and y coordinates (Fig. 3.13). That is $\mathbf{r} = \langle r_x, r_y \rangle$ and $\mathbf{s} = \langle s_x, s_y \rangle$, as illustrated in Fig. 3.15 for $\mathbf{r} = \langle 3, 2 \rangle$ and $\mathbf{s} = \langle 9,8 \rangle$. In this case, the x and y coordinates of points on the straight line segment connecting points $\mathbf{r}$ and $\mathbf{s}$ are given, respectively, by the expressions

$$\lambda \, \mathbf{r}_x + (1 - \lambda) \, \mathbf{s}_x$$
$$\lambda \, \mathbf{r}_y + (1 - \lambda) \, \mathbf{s}_y$$

for values $\lambda \in [0,1]$. Examples for $\lambda = 0.2$ and $\lambda = 0.5$ are shown in Fig. 3.15. For $\lambda = 0.2$, the x coordinate of the point on the segment is

$$0.2 \cdot 3 + 0.8 \cdot 9 = 7.8$$

and its y coordinate is

$$0.2 \cdot 2 + 0.8 \cdot 8 = 6.8$$

For $\lambda = 0.5$, the x coordinate is 6 and the y coordinate is 5.

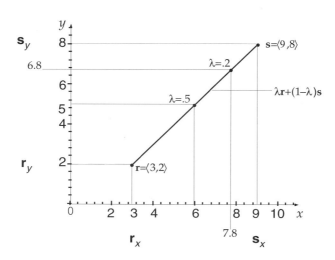

Figure 3.15   The meaning of the expression $\lambda r + (1-\lambda)\mathbf{s}$ in two-dimensional Euclidean space.

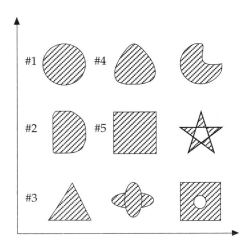

Figure 3.16   Convex and nonconvex sets.

Figure 3.16 shows some convex and nonconvex subsets of the two-dimensional Euclidean space. Examining these subsets, we can easily recognize that sets 1–5 are convex, while the remaining sets are not.

One additional concept, which plays an important role in classical set theory, is the concept of a *partition*. Given a nonempty set A, a family of pairwise disjoint subsets of A is called a partition of A, usually denoted by $\Pi(A)$, if and only if the union of these subsets yields the set A. Formally, a family

$$\Pi(A) = \{A_i \mid i \in I, \varnothing \neq A_i \subseteq A\}$$

of nonempty subsets of A is a partition of A if and only if

$$A_i \cap A_j = \varnothing$$

for each pair $i \neq j$ $(i, j \in I)$ and

$$\bigcup_{i \in I} A_i = A$$

Subsets A_i of a partition are usually called its *blocks*.

For example, the set of students in a class can be partitioned into two subsets (blocks), male students and female students. Alternatively, they can be partitioned by their grades in the course; each block of this partition consists of students with the same grade.

A partition $\Pi_1(A)$ is called a *refinement* of another partition $\Pi_2(A)$ of the same set A if and only if every block of $\Pi_1(A)$ is contained in a block of $\Pi_2(A)$. For example, the partition of a set of students into male students and female students can be refined by partitioning each block into smaller blocks for work on a class project.

EXERCISES

3.1 Which of the following pairs of sets represent equal sets?
  (a) $A = \{0, 1, 2, 3\}, B = \{1, 3, 2, 0\}$
  (b) $A = \{0, 1, 0, 2, 3\}, B = \{0, 1, 2, 3, 2\}$
  (c) $A = \{0, 1, 2, 3, 4\}, B = \{0, 1, 2, 3\}$
  (d) $A = \{x \mid x \text{ is an integer}, 4 < x < 12\}, B = \{5, 6, 7, 8, 9, 10, 11\}$
  (e) $A = \{x \mid x \text{ is } 0, 1, 2 \text{ or } x \text{ is a natural number divisible by } 3, x < 30\}$.
       $B = \{0, 1, 2, 3, 6, 9, 12, 15, 18, 21, 24, 27\}$
  (f) $A = \varnothing, B = \{\varnothing\}$
  (g) $A = \{0\}, B = \{\varnothing\}$
3.2 Which of the following definitions are acceptable as definitions of sets?
  (a) $A = \{a \mid a \text{ is a real number}\}$
  (b) $B = \{b \mid b \text{ is a real number much greater than } 2\}$
  (c) $C = \{c \mid c \text{ is a living organism}\}$
  (d) $D = \{d \mid d \text{ is a decimal digit}\}$
  (e) $E = \{e \mid e \text{ is an exercise in this book}\}$
  (f) $F = \{f \mid f \text{ is a set}\}$
  (g) $G = \{g \mid g \text{ is an intelligent girl}\}$
  (h) $H = \{h \mid h \text{ is a set}, h \notin h\}$
  (i) $I = \{i \mid i \text{ is a subset of } I\}$
  (j) $J = \{j \mid j \text{ is a partition of } J\}$
3.3 Which of the following sets is the empty set $\varnothing$?
  (a) $A = \{a \mid a < 3, a > 3\}$
  (b) $B = \{b \mid b < 3 \text{ or } b > 3\}$
  (c) $C = \{c \mid c = 2, c \text{ is an irrational number}\}$
  (d) $D = \{d \mid d \text{ is a proper subset of } X, X = \{x\}\}$
  (e) $E = \{e \mid e \text{ is a positive integer}, e < 1\}$
3.4 Which of the following statements are correct provided that $X = \{\varnothing, 1, 2, 3, \{2, 3\}\}$?
  (a) $\{2, 3\} \in X$ **(f)** $\{0, 1, 2, 3, \{2, 3\}\} = X$
  (b) $\{2, 3\} \subset X$ **(g)** $\{1, 2, 3, \{2, 3\}\} = X$
  (c) $\varnothing \subset \{2, 3\}$ **(h)** $\{\{2, 3\}\} \subset X$

(d) $\varnothing \subset \varnothing$ (i) $\{1, 3\} \in X$

(e) $\{\varnothing, 0, 1\} \subset X$ (j) $\{\{1, 3\}\} \subset X$

3.5 Let $A = \{a \mid a$ is a rational number$\}$, $B = \{b \mid b$ is a irrational number$\}$, $C = \{0, 1\}$, $D = \{d \mid d$ is an integer, $1 < d < 10\}$, $E = \{e \mid e$ is a decimal digit$\}$. Determine the following sets:

(a) $A - B, C - A, E - D$

(b) $A \cap B, A \cup B, A \cap C, B \cap C, C \cup D$

(c) $(C \cap D) \cup B, (A \cup B) \cap E, (D \cap E) \cap A$

(d) set of all partitions of C

(e) all pairs of the sets A, B, C, D, E that are disjoint

(f) Cartesian products $C \times C$, $C \times C \times C$, and $C \times E$

(g) the power set of $C \cup \{2,3\}$

3.6 Using the properties in Table 3.1, show that the following equalities hold for any sets A and B:

(a) $A \cup (\overline{A} \cap B) = A \cup B$

(b) $A \cap (\overline{A} \cup B) = A \cap B$

(c) $\overline{A \cap \overline{B}} = B \cup (\overline{A} \cap \overline{B})$

(d) $A \cap \overline{B}) \cup (\overline{A} \cap B) = \overline{(\overline{A} \cup B) \cap (A \cup \overline{B}}$

3.7 Show that the equalities in Exercise 3.6 hold by using characteristic functions of the sets involved and Eqs. (3.3)–(3.5).

3.8 For each equality in Exercise 3.6, determine its dual equality.

3.9 Which of the following sets A, B, C defined on the one-dimensional or two-dimensional Euclidean spaces by their characteristic function are convex?

(a) $\chi_A(x) = \begin{cases} 1 & \text{when } x = 0 \text{ or } x \in [1, 2] \\ 0 & \text{otherwise} \end{cases}$

(b) $\chi_B(x, y) = \begin{cases} 1 & \text{when } x \geq 0 \text{ and } 0 \leq y \leq \sin x \\ 0 & \text{otherwise} \end{cases}$

(c) $\chi_C(x, y) = \begin{cases} 1 & \text{when } x \geq 0 \text{ and } 0 \leq y \leq x \\ 0 & \text{otherwise} \end{cases}$

3.10 Which of the following families of subsets of $X = \{0, 1, \ldots, 9\}$ are partitions of X?

(a) $\{\{0, 1\}, \{2, 3, 4, 5, 6, 7, 8, 9\}, X\}$

(b) $\{\{0, 1, 2, 3, 4\}, \{5, 6, 7, 8, 9\}\}$

(c) $\{\varnothing, X\}$

(d) $\{X\}$

(e) $\{\{0,1\}, \{2, 3\}, \{4, 5\}, \{6, 7\}, \{9\}\}$

(f) $\{\{0, 9\}, \{1, 8\}, \{2, 7\}, \{3, 6\}, \{4, 5\}\}$

3.11 Make a list of all partitions of the set $\{1, 2, 3\}$ and determine for each pair of the partitions whether one of the partitions is a refinement of the other one.

4

Fuzzy Sets: Basic Concepts and Properties

4.1 RESTRICTIONS OF CLASSICAL SET THEORY AND LOGIC

The overview of classical set theory in the previous chapter emphasizes one of its central assumptions: The boundaries of classical sets are required to be drawn precisely and, therefore, set membership is determined with complete certainty. An individual is either definitely a member of the set or definitely not a member of it. This sharp distinction is also reflected in classical logic, where each proposition is treated as either definitely true or definitely false.

However, most sets and propositions are not so neatly characterized. For example, the set of tall people is a set whose exact boundary cannot be precisely determined. In the case of the coins, we can determine the members of the set of legal U. S. coin types at any given time by simple enumeration and inspection. However, no amount of simple inspection will be sufficient to answer the question of whether or not an individual belongs in the set of tall people. This is because there is a continuous transition in the height of a person from being *not tall* to being *tall*. A line drawn at any specific height to make the distinction between *tall* and *not tall* is really an artificial demarcation and usually does not represent our concept of who is included in the set of tall people. For example, let us define all tall people as those having a height greater than or equal to 1.8 meters. Then, a person whose height is 1.79 meters will not be considered a tall person. However, according to our intuition, there is no clear distinction between a person of 1.79

meters and a person of 1.8 meters in terms of the word 'tall'. If a job announcement stated that the only requirement for the job was to be tall and one applicant, with height 1.80 meters, received an offer, then an unsuccessful applicant, with height 1.79 meters, would probably feel he had been unfairly excluded from the advertised position.

In effect, the unfair distinction in this example may best be exposed by showing that the distinction between *tall* and *not tall* cannot be properly represented by two sets with precise boundaries. There are many examples of these kinds of concepts that cannot be represented fully by classical sets. They include, for instance, concepts such as *high salary, populous city, accurate clock, coast line, extensive administrative experience*, and so on. Fuzzy sets are a natural tool to characterize such concepts.

Examining the various properties of crisp sets and the relationships crisp sets bear to each other, we can see that the meanings of some of these properties depend particularly on the feature of sharpness of set boundaries. This is especially pronounced in the case of two important laws of both classical set theory and classical logic: the law of contradiction and the law of the excluded middle.

The law of contradiction says that any proposition affirming a fact and denying it at the same time is false. In set-theoretic terms, it says that the same individual cannot simultaneously be a member of a set and its complement. The law of the excluded middle is closely related: It says that any proposition must be either true or false, but not both. Similarly, in classical set theory, it says that an individual must be a member of either a set or its complement.

In our daily lives, however, the stipulation that a statement be either true or false usually does not hold. For example, the statement "John is healthy" cannot be evaluated simply by a definite yes or no. Unless we know that John is a triathlete or the victim of a serious illness, we are prepared to say only that such a sentence is more or less true, because we do not have any strict criteria for the clean demarkation between healthy and not healthy. Similarly, a set of "healthy people" is allowed in classical set theory only if significant, simplifying assumptions are made and the partition between healthy people and unhealthy people is imposed.

Without the imposition of arbitrary partitions, these kinds of sets are ruled out in classical set theory, a circumstance that has prevented classical mathematics from functioning fully in disciplines dealing with vagueness and other kinds of uncertainty. For example, this has created great difficulties in the area of artificial intelligence. The primary aim of this field is to replicate human decision making in a machine, such as a computer or a robot. But virtually all human activities involve reasoning based on vague concepts and incomplete information. For example, our experience says "cliffs are dangerous." Human beings can easily identify a cliff, but how can we create a robot that has the ability of human beings to identify a cliff? The classical yes or no logic is one of many obstacles in this enterprise.

If we allow sets to have imprecise boundaries, then the two classically important principles—the laws of contradiction and excluded middle—will no longer always be true. We might be able to allow that the same person be considered "tall" to some degree and be considered "not tall" to another degree. This relaxation has, as might be expected, deep consequences for the methods with which we apply logic and set theory to the solutions of complex problems. Indeed, it plays a key role in bridging the gap between imprecise concepts which we use to describe reality, and precise classical mathematics.

4.2   MEMBERSHIP FUNCTIONS

As already mentioned, one of the principal motivations for introducing fuzzy sets is to represent imprecise concepts. Because an individual's membership in a fuzzy set may admit some uncertainty, we say that its membership is a matter of degree. Accordingly, a person is a member of the set "tall people" to the degree to which he or she meets the operating concept of "tall." Alternatively, we can say that the *degree of membership* of an individual in a fuzzy set expresses the *degree of compatibility* of the individual with the concept represented by the fuzzy set.

Each fuzzy set, A, is defined in terms of a relevant universal set, X, by a function analogous to the characteristic function introduced in Sec. 3.4. This function, called a *membership function*, assigns to each element x of X a number, $A(x)$, in the closed unit interval [0,1] that characterizes the degree of membership of x in A. Membership functions are thus functions of the form

$$A: X \to [0,1]$$

as depicted in Fig. 4.1. In defining a membership function, the universal set X is always assumed to be a classical set.

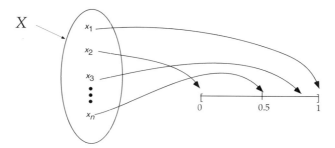

Figure 4.1   General form of membership functions.

Observe that, according to the introduced notation, the same symbol is used for a fuzzy set and its membership function. Since each fuzzy set is uniquely defined by one particular membership function, no ambiguity results from this double use of the symbol.

Contrary to the qualitative symbolic role of numbers 1 and 0 in characteristic functions of classical sets, numbers involved in membership functions of fuzzy sets have a quantitative meaning. This meaning can be extended to classical sets as well, provided they are viewed as special fuzzy sets and their characteristic functions are viewed as special membership functions. These special fuzzy sets, which from this point of view coincide with classical sets, are usually referred to as *crisp sets*.

As examples, let us consider the two membership functions in Fig. 4.2, both expressed in terms of their graphs. These membership functions define two sets of people by their ages. The first one defines the set of teenagers, which is a crisp set; the second one defines a set of young people, as perceived in a particular context.

4.3 REPRESENTATIONS
OF MEMBERSHIP FUNCTIONS

As explained in Sec. 4.2, each fuzzy set is uniquely defined by a membership function. In this section, we introduce the most common ways in which membership functions are represented.

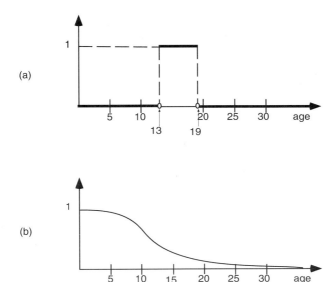

Figure 4.2   Membership function of (a) the set of teenagers; (b) the set of young people.

Graphical Representation

Graphical representation, as exemplified in Fig. 4.2, seems to be the one used most frequently in the literature, perhaps due to its intuitive appeal. Let us elaborate on this representation a little more by looking at some additional examples.

Consider three fuzzy sets defined within a universal set that consists of the seven levels of education defined in Table 4.1. The definitions are precise. They are applied to a person in accordance with his or her educational record. These terms are often used on employment application forms. However, in daily life, we sometimes just refer to someone as *highly educated, little educated*, and so on. As we have seen, vague concepts such as these cannot be formulated in classical set theory, but they can be well formulated by fuzzy sets.

For example, the condition *very highly educated* can be represented by a fuzzy set to which some of the seven levels of education belong to a high degree—levels 5 and 6, for example—whereas the

TABLE 4.1 LEVELS OF EDUCATION

Level Number	Educational Level Attained
0	no education
1	elementary school
2	high school
3	two-year college degree
4	bachelor's degree
5	master's degree
6	doctoral degree

other levels belong to that fuzzy set to a lesser degree. Also, the condition "little educated" can be represented by a different fuzzy set to which the seven educational levels are members to varying degrees. Membership functions of three fuzzy sets, which attempt to capture the concepts described in the natural-language expressions *little educated*, *highly educated*, and *very highly educated*, are represented in Fig. 4.3.

According to this representation, the educational level of a person with a bachelor's degree is compatible with the concept *highly educated* to the degree of 0.8, but compatible with the concept *very highly educated* only to the degree of 0.5. This illustrates that a person with a bachelor's degree is considered to possess a significant amount of education; however, since we know that this degree is surpassed by two graduate degrees, we consider that the label *highly educated* fits the holder of the bachelor's degree rather well, but that the label *very highly educated* is less appropriate.

Let us again suppose that a university defines class levels according to Table 4.2 and that we are seeking to represent the concept of an experienced undergraduate student. By contrast with the crisp sets based on the precisely defined class levels, the vague term *experienced undergraduate student* corresponds to a genuine fuzzy set. This fuzzy set consists of individuals whose degrees of membership in the set range from 0 to 1, and thus the graph of their membership degrees provides a transition from 0 to 1. Depending on our judgment of how many completed credit hours are required for an undergraduate student to be regarded as *experienced*, we might represent the transition from inexperienced to experienced as depicted in Fig. 4.4.

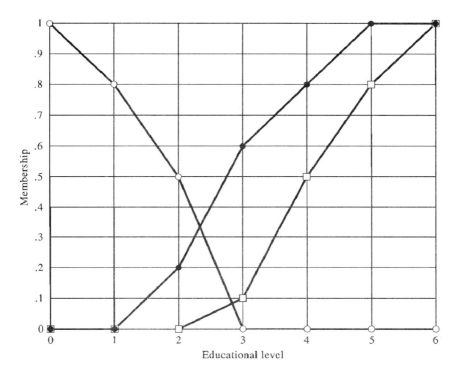

Figure 4.3 Examples of fuzzy sets expressing the concepts of people who are *little educated* (○), *highly educated* (•), and *very highly educated*(❑).

TABLE 4.2 UNDERGRADUATE
CLASS LEVELS

Class Level	Credit Hours
Freshman	0–32
Sophomore	33–62
Junior	63–94
Senior	95–126

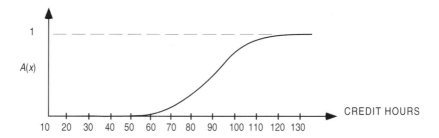

Figure 4.4   Representation of *experienced undergraduate students.*

In this case, the transition from nonmembership to full membership takes the form of a smooth curve which increases in height from left to right. Membership functions may exhibit other types of shapes, such as the shape of the membership function of *young* in Fig. 4.2, which decreases in height from left to right. Further, the membership function for the fuzzy concept *middle-aged* should have something of a bell shape, as shown in Fig. 4.5.

However, in many fuzzy sets the exact shape of the transition from 0 to 1 is not critical. Indeed, sometimes we do not know for sure how to draw the transition from zero membership to total membership to capture the meaning of a linguistic term, such as *medium*, in a given context. The reason is that such a transitional shape must be based on empirical evidence of how the term in question is used in that context; many times this evidence is incomplete. However, most applications of fuzzy set theory do not show great sensitivity to the actual shapes of the membership functions involved. Hence, simple shapes are usually favored.

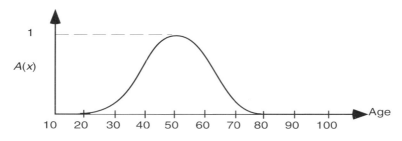

Figure 4.5   Middle-aged person.

Tabular and List Representations

Graphical representation is applicable to membership functions whose universal sets are either the one-dimensional Euclidean space (the real line), as illustrated in the previous section, or the two-dimensional Euclidean space. For other universal sets, alternative representations are needed.

For universal sets that are finite, membership functions can always be represented by tables. The table representing a fuzzy set lists all elements in the universal set and the corresponding membership grades. Using this method, we thus characterize a fuzzy set by a list in which the members of the set are conjoined with their degrees of membership in the set.

Let A be the set of *industrious students in Bio 401*, and let the class of Bio 401 have as members the students Consuelo, Bill, Jiang-Hua, and Kwame. This set A is a subset of the universal set X, *all students at the university*. Unfortunately, not all members of this class study diligently, so some have a low degree of membership in the fuzzy set A, as shown in Table 4.3. Instead of using the table, we may simply list all the ordered pairs consisting of each membership degree together with the label of the individual:

$$A = \{<\text{Consuelo},0.8>, <\text{Bill},0.3>, <\text{Jiang-Hua},0.5>, <\text{Kwame},0.9>\}$$

or

$$A = \{<x_1,0.8>, <x_2,0.3>, <x_3,0.5>, <x_4,0.9>\}$$

This list would often be written in the literature in the following way:

$$A = 0.8/\text{Consuelo} + 0.3/\text{Bill} + 0.5/\text{Jiang-Hua} + 0.9/\text{Kwame}$$

Notice that the symbol / does not stand for division here. It stands for the correspondence between an element in the universal set and its membership grade in the fuzzy set. Likewise the symbol + does not stand for summation; it merely connects the elements. Using this notation, the fuzzy set *very highly educated* can be represented in the following form:

$$B = 0/0 + 0/1 + 0/2 + 0.1/3 + 0.5/4 + 0.8/5 + 1/6$$

Usually, however, the elements that have a 0 membership grade are not listed. Thus, the fuzzy set *very highly educated* would be represented by the simpler form:

$$B = 0.1/3 + 0.5/4 + 0.8/5 + 1/6$$

The generalized notation common in the literature has the following form:

$$A = \sum A(x)/x$$

TABLE 4.3   TABULAR REPRESENTATION
OF A MEMBERSHIP FUNCTION

Student (Label)	Degree of Membership in A
Consuelo (x_1)	0.8
Bill (x_2)	0.3
Jiang-Hua (x_3)	0.5
Kwame (x_4)	0.9

Geometric Representation

Membership functions of fuzzy sets can also be represented in geometric terms. To elucidate this form of representation, let the universal set X be a finite set containing n elements:

$$X = \{x_1, x_2, ..., x_n\}$$

Each element of X may be viewed as a coordinate in the n-dimensional Euclidean space. If we restrict values of each coordinate to real numbers in the unit interval [0,1], we obtain a subset of the Euclidean space that is called an *n-dimensional unit cube*. Each point in the unit cube is uniquely characterized by the values of the n coordinates, x_1, x_2, ..., x_n, which are all in [0,1]. If these values are interpreted as membership grades of the individual elements (coordinates) in a fuzzy set, then the point represents the fuzzy set itself. This means that all fuzzy sets that can be defined on a universal set with n elements can be represented by points in the n-dimensional unit cube.

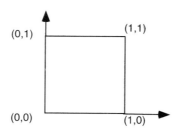

Figure 4.6   Two-dimensional unit cube.

Points in the unit cube with coordinates of 0s or 1s are called *vertices* of the cube. These points represent all crisp sets that can be defined for a universal set with n elements. In the case of $n = 2$, there are only four crisp sets. They are <0, 0>, which is the empty set; <1, 1>, which is the universal set; and <1, 0> and <0, 1>. These four crisp sets are represented by the four vertices of the unit square in two-dimensional space, as shown in Fig. 4.6. Since membership grades of fuzzy sets are numbers between 0 and 1, fuzzy sets are represented by points located anywhere within the square, not merely on its vertices. That is, every fuzzy subset of $X = \{x_1, x_2\}$ is a point covered by the square.

Now, let us look at the case when $n = 3$. Then, there are eight crisp sets: <0, 0, 0>, which is the empty set, <0, 0, 1>, <0, 1, 0>, <0, 1, 1>, <1, 0, 0>, <1, 0, 1>, <1, 1, 0>, as well as <1, 1, 1>, which is the universal set. They correspond to the eight vertices of the unit cube in the three-dimensional space, as shown in Fig. 4.7. Every point of the unit cube represents a fuzzy set.

Analytic Representation

When a universal set is infinite, which is usually the case for a set of real numbers, it is impossible to list all the elements together with their membership grades. For example, the universal set of the fuzzy set *about 6* is the set of all real numbers. This kind of fuzzy set, called a fuzzy number, is often represented by an analytic form, which describes the shape of this fuzzy number. For instance, the fuzzy set

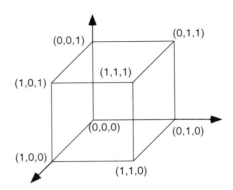

Figure 4.7   Three-dimensional unit cube.

whose graph is shown in Fig. 4.8 may capture for some purpose the concept of *about 6*. It can be expressed in the following analytic form:

$$A(x) = \begin{cases} x - 5 & \text{when } 5 \leq x \leq 6 \\ 7 - x & \text{when } 6 \leq x \leq 7 \\ 0 & \text{otherwise} \end{cases}$$

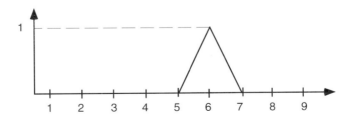

Figure 4.8   A membership function of *about 6*.

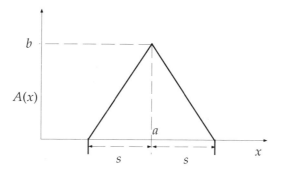

Figure 4.9   Generic, symmetric, and triangular membership functions.

In fact, any symmetric, triangular-shaped membership function which is characterized by the three parameters, a, b, and s, as shown in Fig. 4.9, is represented by the generic form:

$$A(x) = \begin{cases} b\left(1 - \dfrac{|x-a|}{s}\right) & \text{when } a - s \leq x \leq a + s \\[2mm] 0 & \text{otherwise} \end{cases}$$

Another important class of membership functions is trapezoidal shaped, which is captured by the generic graphical representation in Fig. 4.10. Each function in this class is fully characterized by the five parameters, a, b, c, d, and e, via the generic form

$$A(x) = \begin{cases} \dfrac{(a-x)e}{a-b} & \text{when } a \leq x \leq b \\[2mm] e & \text{when } b \leq x \leq c \\[2mm] \dfrac{(d-x)e}{d-c} & \text{when } c \leq x \leq d \\[2mm] 0 & \text{otherwise} \end{cases}$$

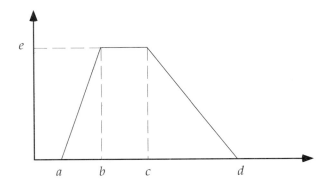

Figure 4.10   Trapezoidal membership function.

Bell-shaped membership functions are also quite common. A generic graph is shown in Fig. 4.11. These functions are represented by the formula

$$A(x) \; = \; ce^{-\dfrac{(x-a)^2}{b}}$$

which involves three parameters, a, b, and c, whose roles are indicated in Fig. 4.11.

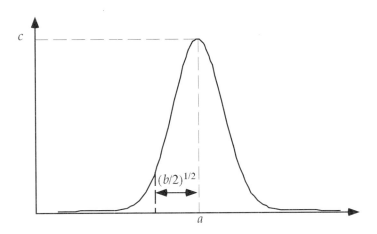

Figure 4.11   Bell-shaped fuzzy set.

4.4   CONSTRUCTING FUZZY SETS

From the previous discussion it is easy to see that membership functions of fuzzy sets play a central role in fuzzy set theory. In each application of fuzzy set theory, we must construct appropriate fuzzy sets (i.e., their membership functions) by which the intended meanings of relevant linguistic terms are adequately captured. These meanings are, of course, strongly dependent on the context in which the terms are used. For example, *young* has a different meaning when applied to children, university professors, or retired people. Its meaning is even more varied when we apply it to different types of objects, such as geological formations, trees, or stars.

The problem of constructing membership functions in the contexts of various applications is not a problem of fuzzy set theory per se. It is a problem of knowledge acquisition, a subject of a relatively new field called knowledge engineering. The process of knowledge acquisition involves one or more experts in the application area and a knowledge engineer. The role of the knowledge engineer is to elicit the knowledge of interest from the experts and to express it in some operational form of a required type.

Many methods for constructing membership functions have been developed, but their full coverage is beyond the scope of this introductory text. In the following, we outline some of the most direct methods.

In some cases, it is reasonable to request that the expert define a membership function for a linguistic term in a given application context, either completely or to exemplify it for some selected individuals in the universal set. To request a complete definition, usually in terms of a justifiable mathematical formula, is not always feasible. It is feasible only for linguistic terms that are perfectly represented in the given application context by some individuals in the universal set, called *ideal prototypes*, provided that the compatibility of other individuals with these ideal prototypes can be expressed mathematically by a meaningful *similarity function.*

For example, let us try to establish the membership function of a fuzzy set E of *almost equilateral triangles.* According to our knowledge of geometry, an equilateral triangle has three equal angles of $60°$, as is depicted in Fig. 4.12a.

Clearly, this triangle, t_1, is an ideal prototype and, hence, $E(t_1) = 1$. Membership grades of other triangles, t_i, in fuzzy set E can be expressed in terms of their closeness to this ideal prototype by comparing the three angles, α_i, β_i, γ_i, of each triangle t_i with the equal angles ($60°$) of the equilateral triangle t_1.

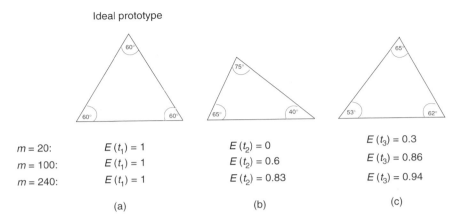

Ideal prototype

$m = 20$: $E(t_1) = 1$ $E(t_2) = 0$ $E(t_3) = 0.3$
$m = 100$: $E(t_1) = 1$ $E(t_2) = 0.6$ $E(t_3) = 0.86$
$m = 240$: $E(t_1) = 1$ $E(t_2) = 0.83$ $E(t_3) = 0.94$

(a)   (b)   (c)

Figure 4.12   Membership grades of the pictured triangles in the fuzzy set of *almost equilateral triangles* under three distinct definitions.

This can be done by the formula

$$
E(t_i) = \begin{cases} 1 - \dfrac{d(t_i)}{m} & \text{when } d(t_i) \le m \\ 0 & \text{when } d(t_i) > m \end{cases}
$$

where $d(t_i) = |\alpha_i - 60| + |\beta_i - 60| + |\gamma_i - 60|$ and m is the largest acceptable deviation (in terms of the three angles) of triangle t_i from the equilateral triangle t_1. Since the largest possible value of $d(t_i)$ is 240, m must be a number chosen from the interval (0, 240]; its choice depends on the purpose for which we define this fuzzy set. The larger the value of m, the more liberal our interpretation of the expression *almost equilateral triangle*. Examples of membership grades calculated by the proposed formula for three triangles and three values of m are shown in Fig. 4.12.

   If it is not feasible to define the membership function in question completely, the expert will often be able to exemplify it for some representative individuals of the universal set. The exemplification may be facilitated by asking the expert questions regarding the compatibility of these representative individuals, denoted by variable x, with the linguistic term that is to be represented by fuzzy set A. The answers to these questions result in a set of pairs $<x, A (x)>$ that exemplify the membership function under construction. This set is then used for constructing the full membership function. One way to do this is to select an appropriate class of functions (triangular, trapezoidal, bell-shaped, etc.) and employ some relevant curve-fitting method to determine the function that fits best with the given samples. Another way is to use an appropriate neural network to construct the membership function by "learning" from the given samples. This approach has been so successful that neural networks are now viewed as a standard tool for constructing membership functions.

When several experts are employed, their opinions must be properly aggregated to determine relevant membership grades. To illustrate this issue, let us use a simple example. Suppose that there are five springboard divers: Alice, Bonnie, Cathy, Dina, Eva. There are also ten referees: $r_1, r_2,..., r_{10}$. We need to determine a membership function A that will capture the linguistic term *excellent diver*. We ask each referee if a particular person is an excellent diver. The answer is required to be either yes or no. Suppose we have the results of our survey listed in Table 4.4, where 1 and 0 stand for yes and no, respectively. Then for every diver we calculate the membership grade of belonging to the fuzzy set A by taking the ratio of the total number of favorable answers to the total number of referees. This procedure results in the following fuzzy set:

$$A = 0.3/\text{Alice} + 0.4/\text{Bonnie} + 0.6/\text{Cathy} + 0.9/\text{Dina} + 0.6/\text{Eva}$$

TABLE 4.4   THE DIVING SURVEY

	Alice	Bonnie	Cathy	Dina	Eva
r_1	1	1	1	1	1
r_2	0	0	1	1	1
r_3	0	1	0	1	0
r_4	1	0	1	1	1
r_5	0	0	1	1	1
r_6	0	1	1	1	1
r_7	0	0	0	0	0
r_8	1	1	1	1	1
r_9	0	0	0	1	0
r_{10}	0	0	0	1	0

4.5 OPERATIONS ON FUZZY SETS

In Sec. 3.2, the three basic operations on classical sets—complement, union, and intersection—were introduced and their fundamental properties were discussed in Sec. 3.3. While these operations are unique in classical set theory, their extensions in fuzzy set theory are not unique. For each of the classical operations there exists a broad class of operations that qualify as their fuzzy generalizations. Distinct operations in each of these classes reflect distinct meanings of the linguistic terms *not*, *and*, and *or* when employed in sentences of natural language in different contexts. However, one special operation in each of the three classes possesses certain desirable properties, which often make it a good approximation of the respective linguistic term. These special operations on fuzzy sets, which are referred to as *standard fuzzy operations*, are by far the most common operations in practical applications of fuzzy set theory. It is thus reasonable to consider, in our exposition of fuzzy set theory in this book, only these standard operations.

Standard Fuzzy Complement

Given a fuzzy set A defined on a universal set X, its complement $\overline{A}$ is another fuzzy set on X that inverts, in some sense, the degrees of membership associated with A. While for each $x \in X$, $A(x)$ expresses the degree to which x *belongs* to A, $\overline{A}(x)$ expresses the degree to which x *does not belong* to A. Using an adaptation of the Venn diagram, an intuitive picture of this idea is shown in Fig. 4.13.

The most natural way to express this idea formally is to use the formula

$$\overline{A}(x) = 1 - A(x) \qquad (4.1)$$

for all $x \in X$. This is indeed how the *standard fuzzy complement* is defined.

To illustrate the meaning of this definition, let us consider the fuzzy set A of *experienced undergraduate students*, whose membership function is given in Fig. 4.4. According to Eq. (4.1), the membership

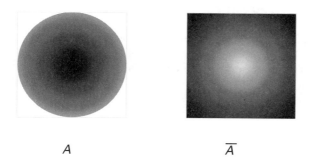

A $\overline{A}$

Figure 4.13   Fuzzy complement.

functions of A and $\overline{A}$ (the set of *inexperienced undergraduate students*) are mirror images of one another, as shown in Fig. 4.14.

One consequence of the imprecise boundaries of fuzzy sets is that they overlap with their complements, as seen from Fig. 4.14. For example, when a student belongs to the set of experienced undergraduate students with the degree of 0.8, then he or she also belongs to the set of inexperienced undergraduate students with a degree of 0.2. This is one of the fundamental differences between classical set theory and fuzzy set theory. In classical set theory, sets never overlap with their complements.

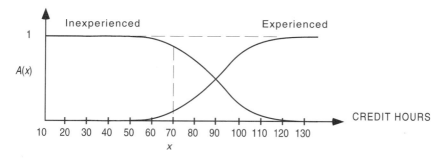

Figure 4.14   The set of *experienced undergraduate students* and its complement.

Standard Fuzzy Union

Consider a universal set X and two fuzzy sets A and B defined on X. Then the *standard fuzzy union* of A and B, denoted by $A \cup B$, is defined by the membership functions via the formula

$$(A \cup B)(x) = \max [A(x), B(x)] \tag{4.2}$$

for all $x \in X$.

As an example, let X be a set of n doctor's patients identified by numbers 1, 2, ..., n, and let A and B denote fuzzy sets of those patients in X that have *high blood pressure* and *high fever*, respectively. Then, using Eq. (4.2), we can determine the set $A \cup B$ of patients in X that have *high blood pressure or high fever*, as illustrated in Table 4.5.

TABLE 4.5   ILLUSTRATION OF THE STANDARD FUZZY UNION

Patients	A=high blood pressure	B=high fever	A∪B
1	1.0	1.0	1.0
2	0.5	0.6	0.6
3	1.0	0.1	1.0
...	...	...	...
n	0.1	0.7	0.7

Another example is the union of the set of *experienced undergraduate students* and its complement, the set of *inexperienced undergraduate students*, shown in Fig. 4.15 by the bold line over the shaded area. This example illustrates that the *law of excluded middle*,

$$A \cup \overline{A} = X$$

of classical set theory does not hold for fuzzy sets under the standard fuzzy union and the standard fuzzy complement. We can easily see that this law is violated for all elements x of X such that $A(x) \notin \{0, 1\}$. When, for example, $A(x) = 0.6$, we have $\overline{A}(x) = 1 - 0.6 = 0.4$, and

$$(A \cup B)(x) = \max [0.6, 0.4] = 0.6$$

This means that x is not a member of X with full membership and the law of excluded middle is violated.

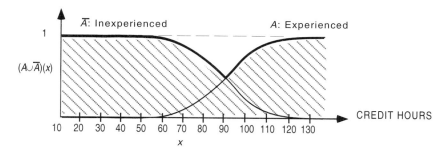

Figure 4.15 Union of a fuzzy set and its complement.

Standard Fuzzy Intersection

Consider again two fuzzy sets A and B, defined on X. The *standard fuzzy intersection*, denoted $A \cap B$, is defined by the membership functions via the formula

$$(A \cap B)(x) = \min\ [A(x),\, B(x)] \tag{4.3}$$

for all $x \in X$.

Let A denote a fuzzy set of rivers that are *long* and let B denote a fuzzy set of rivers that are *navigable*. Then $A \cap B$ is a fuzzy set of rivers that are *long and navigable*. This is illustrated in Table 4.6 for a sample of five rivers.

TABLE 4.6 ILLUSTRATION OF THE STANDARD FUZZY INTERSECTION

River	A = Long River	B = Navigable River	$A \cap B$
Amazon	1.0	0.8	0.8
Nile	0.9	0.7	0.7
Yang-Tse	0.8	0.8	0.8
Danube	0.5	0.6	0.5
Rhine	0.4	0.3	0.3

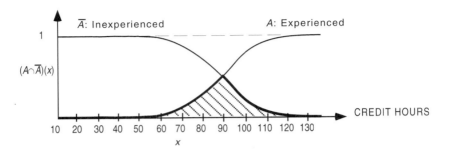

Figure 4.16   Intersection of a fuzzy set and its complement.

Another example is the intersection of the set of *experienced undergraduate students* and its complement as shown in Fig. 4.16. This example illustrates that the *law of contradiction*

$$A \cap \overline{A} = \varnothing$$

of classical set theory does not hold for fuzzy sets under the standard fuzzy intersection and standard fuzzy complement. When, for example, $A(x) = 0.6$ for some $x \in X$, then $\overline{A}(x) = 1 - 0.6 = 0.4$ and

$$(A \cap B)(x) = \min [A(x), \overline{A}(x)] = 0.4$$

That is, x is a member of $A \cap \overline{A}$ with the degree of 0.4 and not with the degree of 0 as demanded by the law of contradiction.

Fundamental Properties

As already shown, the standard fuzzy operations do not satisfy two laws of their classical counterparts: the law of excluded middle and the law of contradiction. This is a consequence of the imprecise boundaries of fuzzy sets.

It can easily be verified that the standard fuzzy operations satisfy all other properties of the corresponding operations in classical set theory, as discussed in Sec. 3.3. These are commutativity, associativity, idempotency, distributivity, and the De Morgan laws. Other nonstandard fuzzy operations do not satisfy all of these properties. For example, the standard operations are the only ones that are idempotent.

By restricting ourselves to the standard fuzzy operations, the great expressive power of fuzzy set theory is not fully utilized. In particular, the standard fuzzy operations are not capable of expressing the full variety of meanings of the linguistic terms *or*, *and*, and *not* when applied to fuzzy concepts of natural language. However, the standard fuzzy operations have been found adequate in most practical applications of fuzzy set theory. Consequently this restriction is not severe, at least for the purpose of this introductory text.

EXERCISES

4.1 Read some articles in a newspaper and identify all linguistic expressions that are inherently fuzzy. For each of the identified expressions, try to define a reasonable fuzzy set to represent it.

4.2 Is the membership function of a fuzzy set unique? Explain your answer.

4.3 Why are crisp sets special cases of fuzzy sets? What is the significance of this?

4.4 Within the range [0,100], define membership functions of fuzzy sets *young, middle age, old* according to your intuition. Then
(a) Plot your membership functions.
(b) Determine the standard fuzzy complements of these three sets.
(c) Determine the standard intersections and unions for each pair of the three sets.

4.5 Why is the law of excluded middle violated in fuzzy set theory when the standard set operations are employed? What is the significance of this?

4.6 Let fuzzy sets A and B be defined by the following membership functions:

$$A(x) = \begin{cases} 1 - \dfrac{|x-3|}{2} & \text{when } 1 \leq x \leq 5 \\ 0 & \text{otherwise} \end{cases}$$

$$B(x) = \begin{cases} 1 - \dfrac{|x-4|}{2} & \text{when } 2 \leq x \leq 6 \\ 0 & \text{otherwise} \end{cases}$$

(a) Plot these membership functions.
(b) Determine analytic expressions and plots of $\overline{A}$ and $\overline{B}$.
(c) Determine analytic expressions and plots of $A \cap B$, $A \cup B, \overline{A} \cap \overline{B}$ and $\overline{A} \cup \overline{B}$.

4.7 Let X be the set of all rectangles and let each rectangle be characterized by the lengths of its adjacent sides, denoted by a and b. Employing the ratio a/b, explore possible definitions of fuzzy sets of approximate squares.

4.8 Determine the generic form for triangular-shaped membership functions that are not required to be symmetric.

5

FUZZY SETS:
FURTHER PROPERTIES

Throughout previous chapters, the explication of the concept of the fuzzy set centers on drawing important distinctions between fuzzy sets and classical sets. However, there are also important ways in which classical sets and fuzzy sets are connected and these are explored in this chapter.

5.1 α-CUTS OF FUZZY SETS

In Sec. 4.3, several ways in which we may represent fuzzy sets were introduced. In all these representations, each member x of the universal set X is assigned a unique membership degree $A(x)$ in the represented set A. The representations differ from each other in the way in which these assignments are expressed: graphs, tables, lists, mathematical formulas, or coordinates in the n-dimensional unit cube. They can be viewed as representations of the same type.

There are also representations of fuzzy sets of another type based on specific assignments of numbers in [0,1] to crisp sets. To facilitate our explanation of these representations, let us consider a fuzzy set E intended to capture the concept of an *expensive book* as conceived by a particular buyer in a given context (certain category of books, country, etc.). For the sake of simplicity, let the price range (i.e., our universal set) be from 0 (for books that are free) to $100, and let

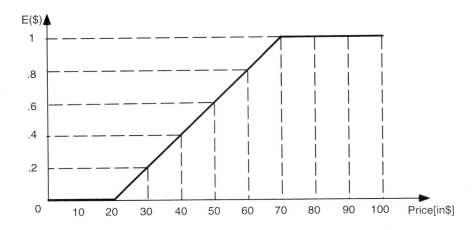

Figure 5.1   Fuzzy set expressing the concept of an expensive book.

the membership function of E be defined by the graph in Fig. 5.1. We may also consider a discrete version of the membership function, in which $X = \{0, 10, 20, \ldots, 100\}$ and $E = 0/0 + 0/10 + 0/20 + 0.2/30 + 0.4/40 + 0.6/50 + 0.8/60 + 1/70 + 1/80 + 1/90 + 1/100$.

A given fuzzy set X is always associated with a family of crisp subsets of X. Each of these subsets consists of all elements of X whose membership degrees in the fuzzy set are restricted to some given crisp subset of [0,1]. For example, considering the introduced fuzzy set E, we can find from the graph in Fig. 5.1 the price range of books that belong to this set with a degree in the closed interval [0.2,0.6]. It is expressed by the closed interval [30, 50], which is a crisp subset of the universal set [0,100]. Similarly, the price range of books that belong to the fuzzy set with a degree of 0.8 or less is [0,60], and those that belong to it with a degree of 0.8 or more is [60, 100]. In general, for any restriction of the membership degree we obtain a unique subset of [0,100].

One way of restricting membership degrees is particularly important. It is a restriction of membership degrees that are greater than or equal to some chosen value α in [0,1]. When this restriction is applied to a fuzzy set A we obtain a crisp subset $^{\alpha}A$ of the universal set X, which is called an α-cut (alpha-cut) of A. Formally,

$$^{\alpha}A = \{x \in X \mid A(x) \geq \alpha\}$$

for any $\alpha \in [0,1]$. This equation says that the α-cut of a fuzzy set A is the crisp set $^{\alpha}A$ that contains all the elements of the universal set X whose membership degrees in A are *greater than or equal to* the specified value of α.

For the fuzzy set E (Fig. 5.1), some examples of α-cuts are: $^{0}E = [0, 100]$, $^{0.2}E = [30, 100]$, $^{0.5}E = [45, 100]$, $^{0.9}E = [65, 100]$, $^{1}E = [70, 100]$. For the discrete version of E, the corresponding examples are: $^{0}E = \{0, 10, ..., 100\}$, $^{0.2}E = \{30, 40, ..., 100\}$, $^{0.5}E = \{50, 60, ..., 100\}$, $^{0.9}E = {}^{1}E = \{70, 80, ..., 100\}$. We can see from the examples that by increasing the value of α the size of the α-cut does not increase; it either decreases or remains the same. This is a general property of α-cuts: If we consider different values of α in increasing order, such as 0.1, 0.2, 0.3, and so on, the associated α-cuts are ordered by set inclusion. Formally: For any fuzzy set A, if $\alpha_1 < \alpha_2$, then $^{\alpha_1}A \supseteq {}^{\alpha_2}A$ and, consequently,

$$^{\alpha_1}A \cap {}^{\alpha_2}A = {}^{\alpha_2}A$$

$$^{\alpha_1}A \cup {}^{\alpha_2}A = {}^{\alpha_1}A$$

This means that all α-cuts of any fuzzy set form a family of *nested crisp sets*, as visually depicted in Fig.5.2.

Carrying the concept of the α-cut further, we can define a somewhat more restricted variant, a *strong α-cut*. By definition, it is the crisp set that contains all the elements of the universal set whose

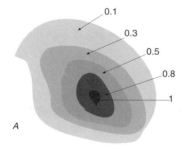

Figure 5.2   Nested structure of α-cuts.

membership grades in the given set are *greater than* (but do not include) the specified value of α. For a fuzzy set A the strong α-cut, $^{\alpha+}A$, is defined in the following way:

$$^{\alpha+}A = \{x \in X \mid A(x) > \alpha\}$$

For the previous example of fuzzy set E, we have: $^{0+}E = (20, 100]$, $^{0.2+}E = (30, 100]$, $^{0.5+}E = (45, 100]$, $^{0.9+}E = (65, 100]$, $^{1+}E = \varnothing$. Counterparts of these strong α-cuts for the discrete version of E are: $^{0+}E = \{30, 40, ..., 100\}$, $^{0.2+}E = \{40, 50, ..., 100\}$, $^{0.5+}E = \{50, 60, ..., 100\}$, $^{0.9+}E = \{70, 80, 90, 100\}$, $^{1+}E = \varnothing$.

Strong α-cuts of fuzzy sets also form families of nested crisp sets: If $\alpha_1 < \alpha_2$, then $^{\alpha_1+}A \supseteq {}^{\alpha_2+}A$, and, consequently,

$$^{\alpha_1+}A \cap {}^{\alpha_2+}A = {}^{\alpha_2+}A$$

$$^{\alpha_1+}A \cup {}^{\alpha_2+}A = {}^{\alpha_1+}A$$

The reason that we place great importance on the α-cuts and strong α-cuts of fuzzy sets is that any fuzzy set may be completely characterized either by its α-cuts or by its strong α-cuts. These characterizations are explained in Sec. 5.2.

Support, Core, and Height

Two of the crisp sets associated with any given fuzzy set A are particularly important. One of them is the set of all elements of the universal set X that have nonzero membership in A. This crisp set, called the *support* of A and usually denoted by supp(A), is defined as the strong α-cut for $\alpha = 0$. That is,

$$\text{supp}(A) = {}^{0+}A = \{x \in X \mid A(x) > 0\}$$

The support of fuzzy set E (Fig. 5.1) is (20, 100].

Another important crisp set connected with any given fuzzy set A is the set of all elements of X for which the degree of membership in A is 1. This set, called the *core* of A and denoted by core(A), is defined in exactly the same way as the α-cut for $\alpha = 1$; hence,

$$\text{core}(A) = {}^{1}A = \{x \in X \mid A(x) \geq 1\} = \{x \in X \mid A(x) = 1\}$$

The core of set E is [70, 100].

The support and core of a fuzzy set are thus particular cases of the strong α-cuts and α-cuts, respectively. Special names and symbols have been introduced for them in the literature because they are frequently used for rough characterization of fuzzy sets.

The core of a given fuzzy set may be empty and, moreover, the α-cuts of a fuzzy set may be empty for all values of α greater than some nonzero value in [0, 1]. The largest value of α for which the α-cut is not empty is called the *height* of a fuzzy set. Alternatively, the height, $h(A)$, of a fuzzy set A may be defined as the largest membership grade obtained by any element in that set. When $h(A) = 1$ set A is called *normal*; otherwise it is called *subnormal*.

The concepts of support, core, height, and α-cuts are illustrated in Fig. 5.3 by a trapezoidal-shaped fuzzy set A defined on $X = [a, b]$. This fuzzy set is normal, as is the fuzzy set E in Fig. 5.1.

Level Set

It is sometimes useful to identify the set of distinct α-cuts of a given fuzzy set A. This can be done by identifying all distinct numbers

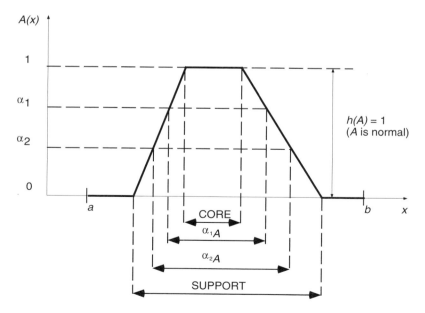

Figure 5.3   Illustration of the concepts of support, core, height, and α-cuts.

in [0,1] that are employed as membership grades of the elements of X in A. This set of numbers, which represents all distinct α-cuts of A, is called the *level* set of A and is denoted by $L(A)$. Formally,

$$L(A) = \{\alpha \in [0, 1] \mid A(x) = \alpha \text{ for some } x \in X\}$$

The level set of fuzzy set E consists of the whole unit interval [0,1]. However, the level set of the discrete version of E is

$$L(E) = \{0, 0.2, 0.4, 0.6, 0.8, 1\}$$

Observe that the α-cut does not change for values of α which are between the numbers in this level set. For example, for any α in the interval $(0,0.2]$, we obtain the same α-cut; similarly, for intervals $(0.2,0.4]$, $(0.4,0.6]$, $(0.6,0.8]$. and $(0.8, 1]$.

5.2 α-CUT REPRESENTATION

As already mentioned, one important role of α-cuts in fuzzy set theory is that they can be used to represent fuzzy sets. Consider the fuzzy set

$$A = 0.2/x_1 + 0.4/x_2 + 0.6/x_3 + 0.8/x_4 + 1/x_5$$

as our example. Its level set is $L(A) = \{0.2, 0.4, 0.6, 0.8, 1\}$. We can represent this fuzzy set by its α-cuts in that we can specify whether or not an element is included in the α-cut defined by some particular value of α. The set A is associated with only five distinct α-cuts, which are defined by the following characteristic functions:

$$^{0.2}A = 1/x_1 + 1/x_2 + 1/x_3 + 1/x_4 + 1/x_5$$
$$^{0.4}A = 0/x_1 + 1/x_2 + 1/x_3 + 1/x_4 + 1/x_5$$
$$^{0.6}A = 0/x_1 + 0/x_2 + 1/x_3 + 1/x_4 + 1/x_5$$
$$^{0.8}A = 0/x_1 + 0/x_2 + 0/x_3 + 1/x_4 + 1/x_5$$
$$^{1}A = 0/x_1 + 0/x_2 + 0/x_3 + 0/x_4 + 1/x_5$$

We now convert each of the α-cuts to a special fuzzy set, $_\alpha A$, defined for each $x \in X$ by the formula

$$_\alpha A(x) = \alpha{}^\alpha A(x) \tag{5.1}$$

Applying this formula to each α-cut of our fuzzy set A, we obtain the following results:

$$_{0.2}A = 0.2/x_1 + 0.2/x_2 + 0.2/x_3 + 0.2/x_4 + 0.2/x_5$$
$$_{0.4}A = 0/x_1 + 0.4/x_2 + 0.4/x_3 + 0.4/x_4 + 0.4/x_5$$
$$_{0.6}A = 0/x_1 + 0/x_2 + 0.6/x_3 + 0.6/x_4 + 0.6/x_5$$
$$_{0.8}A = 0/x_1 + 0/x_2 + 0/x_3 + 0.8/x_4 + 0.8/x_5$$
$$_{1}A = 0/x_1 + 0/x_2 + 0/x_3 + 0/x_4 + 1/x_5$$

It is now easy to see that the union of these five special fuzzy sets is exactly the original fuzzy set A. That is,

$$A = {}_{0.2}A \cup {}_{0.4}A \cup {}_{0.6}A \cup {}_{0.8}A \cup {}_{1}A$$

This property holds generally and is stated in the following theorem.

Theorem 5.1. For any $A \in \mathcal{F}(X)$,

$$A = \bigcup_{\alpha \in [0,1]} {}_\alpha A \tag{5.2}$$

where $_\alpha A$ is defined by (5.1) and $\cup$ denotes the standard fuzzy union.

This important theorem is sometimes called a *decomposition theorem of fuzzy sets*. We should also mention that a similar theorem is obtained when $_\alpha A$ is based on strong α-cuts, that is, when ${}^\alpha A$ in Eq. (5.1) is replaced with ${}^{\alpha+}A$.

5.3   CUTWORTHY PROPERTIES OF FUZZY SETS

The α-cut representation of fuzzy sets introduces an important connection between classical set theory and fuzzy set theory. This connection allows us to extend the various properties of classical sets to their fuzzy-set counterparts. A property of classical sets is extended to fuzzy sets via the α-cut representation by requiring that the property be satisfied (in the classical sense) in all α-cuts of the relevant fuzzy sets. Any property of fuzzy sets that is derived in this way from classical set theory is called a *cutworthy property*. In this section, we examine some simple cutworthy properties of fuzzy sets. Other, more complicated cutworthy properties are introduced and discussed in later chapters.

One of the simplest examples of a cutworthy property is the equality of two fuzzy sets, A and B, defined on the same universal set X. The usual definition of this equality is

$$A = B \text{ if and only if } A(x) = B(x) \text{ for all } x \in X$$

However, there exists an alternative but equivalent definition of equality, which is based on the α-cut representations of the fuzzy sets involved:

$$A = B \text{ if and only if } {}^{\alpha}A = {}^{\alpha}B \text{ for all } \alpha \in [0, 1]$$

Therefore, the equality of fuzzy sets is a cutworthy property. Let us prove that the two definitions of equality of fuzzy sets are equivalent.

It is clear that if $A = B$, then ${}^{\alpha}A = {}^{\alpha}B$ for any $\alpha \in [0,1]$. Conversely, if ${}^{\alpha}A = {}^{\alpha}B$ for all $\alpha \in [0,1]$, A must be equal to B. Let us suppose $A \neq B$, that is, there is an element x in the universal set such that its membership grade in set A is not equal to its membership grade in B: $A(x) \neq B(x)$. Then either $A(x) > B(x)$ or $A(x) < B(x)$. If $A(x) > B(x)$ and $\alpha = A(x)$, then $x \in {}^{\alpha}A$, but $x \notin {}^{\alpha}B$. This contradicts the fact that ${}^{\alpha}A = {}^{\alpha}B$ for any $\alpha \in [0,1]$. If $A(x) < B(x)$ we arrive at the same contradictory result.

In a similar way we can prove that $A \subseteq B$ if and only if ${}^{\alpha}A \subseteq {}^{\alpha}B$ for any $\alpha \in [0,1]$. Therefore, the *set inclusion* of fuzzy sets is also a cutworthy property. The standard fuzzy union and intersection are also cutworthy, as expressed by the following theorem.

Theorem 5.2. For any two fuzzy sets A, B and $\alpha \in [0,1]$,

(a) ${}^{\alpha}(A \cup B) = {}^{\alpha}A \cup {}^{\alpha}B$

(b) ${}^{\alpha}(A \cap B) = {}^{\alpha}A \cap {}^{\alpha}B$

The meaning of this theorem is illustrated in Fig. 5.4. To illustrate the theorem by a specific numerical example, let

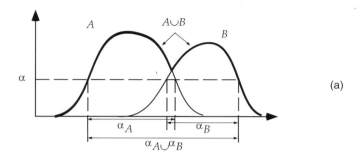

(a)

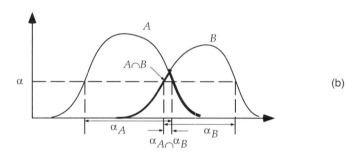

(b)

Figure 5.4   Illustration of Theorem 5.1.

$$A = 0.2/x_1+0.4/x_2+0.6/x_3+0.8/x_4+1/x_5$$
$$B = 1/x_1+0.7/x_2+0.5/x_3+0.3/x_4+0.1/x_5$$

Then,

$$A\cup B=1/x_1+0.7/x_2+0.6/x_3+0.8/x_4+1/x_5$$
$$A\cap B=0.2/x_1+0.4/x_2+0.5/x_3+0.3/x_4+0.1/x_5$$

Now, let $\alpha=0.5$; then, $^{\alpha}A=\{x_3, x_4, x_5\}$, $^{\alpha}B=\{x_1, x_2, x_3\}$, $^{\alpha}(A\cup B)=\{x_1, x_2, x_3, x_4, x_5\}$ and $^{\alpha}(A\cap B)=\{x_3\}$. Similarly, $^{\alpha}A\cup^{\alpha}B =\{x_1, x_2, x_3, x_4, x_5\}= ^{\alpha}(A\cup B)$, $^{\alpha}A\cap^{\alpha}B = \{x_3\} = ^{\alpha}(A\cap B)$.

Our last example of a cutworthy property in this section is the *convexity of fuzzy sets* when defined via their α-cut representations. A fuzzy set defined on the set of real numbers (or, more generally, on any

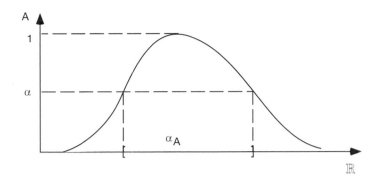

Figure 5.5 A convex fuzzy set of real numbers.

n-dimensional Euclidean space) is said to be convex if and only if all of its α-cuts are convex in the classical sense (as introduced in Sec. 3.5).

By inspecting the graph of the membership function of a given fuzzy set defined on the real line, it is quite easy to determine whether the fuzzy set is convex or not. For a fuzzy set to be convex the graph must have just one peak. A typical graph of a convex fuzzy set is given in Fig. 5.5. Although only one α-cut is shown in this figure, it is easy to see that all α-cuts in this example are closed intervals of real numbers and, hence, convex sets in the classical sense. The membership function of a fuzzy set that is not convex is exemplified in Fig. 5.6. The α-cut shown in the figure illustrates the fact that some α-cuts are not convex in this example.

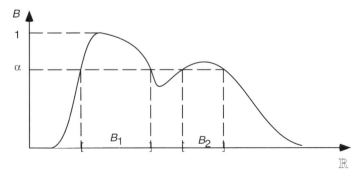

Figure 5.6 Nonconvex fuzzy set ($^{\alpha}B = B_1 \cup B_2$).

5.4   EXTENSION PRINCIPLE

Having examined the similarities and differences between the realm of crisp sets and that of fuzzy sets, we are naturally led to the consideration that in order to develop computation with fuzzy sets we need to find a way to take traditional, crisp functions and fuzzify them. A principle for fuzzifying crisp functions is called the *extension principle*.

To understand this principle let us examine the following problem. Suppose that in a company the distribution of employees' ages and their salaries is as described in Table 5.1. Now, the question we pose is: What is a young employee's salary? To answer this question, we first regard the table as a function f from the set $X=\{20, 25, 30, 35, 40, 45, 50, 55, 60, 65\}$ to the set $Y=\{2.5, 3, 3.5, 4, 4.5, 5\}$. The first step in our procedure is to formulate the meaning of the concept *young* as a fuzzy set A of general form $A = A(x)/x$ by establishing membership grades for all $x \in X$. Assume that the intended meaning is well expressed in the given context by the fuzzy set $A=1/20+1/25+0.8/30+0.6/35+0.4/40+0.2/45+0/50+0/55+0/60+0/65$, whose graph is shown in Fig. 5.7. Using this fuzzy set and the information given in Table 5.1, we need to determine an appropriate fuzzy set B that captures the meaning of the linguistic expression *young employee's salary*. This fuzzy set is dependent on A via function f which for each x in X assigns a particular $y = f(x)$ in Y according to Table 5.1. This dependence is expressed by the general form

$$B = A(x)/f(x)$$

For A and f given in our example we have

$$B = 1/f(20) + 1/f(25) + 0.8/f(30) + 0.6/f(35) + 0.4/f(40) + 0.2/f(45) \\ + 0/f(50) + 0/f(55) + 0/f(60) + 0/f(65) = 1/2.5 + 1/2.5 + 0.8/3 \\ + 0.6/3.5 + 0.4/3.5 + 0.2/4 + 0/4 + 0/4.5 + 0/4.5 + 0/5$$

This expression matches the membership grade of each age in the concept *young* to the corresponding salary at that age.

TABLE 5.1   EMPLOYEES AND THEIR SALARIES

Age (in years)	20	25	30	35	40	45	50	55	60	65
Salary ($ in K)	2.5	2.5	3.0	3.5	3.5	4.0	4.0	4.5	4.5	5.0

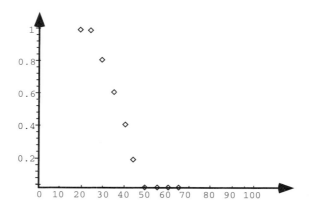

Figure 5.7   Fuzzy set employed to express the concept of a *young employee*.

Observe that some salaries are associated (according to function f defined by Table 5.1) with more than one age. For example, the salary of \$3.5K is associated with the ages 35 and 40. As a consequence, two distinct membership grades (0.6 and 0.4) are assigned to this salary in the expression for B. Due to the disjunctive nature of the association of salaries with ages (salary of \$3.5K is associated with the ages of 35 or 40), we take the maximum of the two membership grades. This results in the final expression

$$B = 1/2.5 + 0.8/3 + 0.6/3.5 + 0.2/4 + 0/4.5 + 0/5$$

which defines the salary of young employees in the company. The graph of B is shown in Fig. 5.8.

The described example, which is summarized in Fig. 5.9, illustrates the use of the extension principle. In more general terms, but restricted to finite universal sets, the principle is stated as follows.

Extension Principle. Let $f: X \to Y$, where X and Y denote finite crisp sets, be a given function. Two functions may be induced from f. One, denoted by $\tilde{f}$, is a function from $\mathcal{F}(X)$ to $\mathcal{F}(Y)$. The other, denoted by $\tilde{f}^{-1}$, is a function from $\mathcal{F}(Y)$ to $\mathcal{F}(X)$. These functions are defined by

$$[\tilde{f}(A)](y) = \max_{x \mid y = f(x)} A(x) \tag{5.3}$$

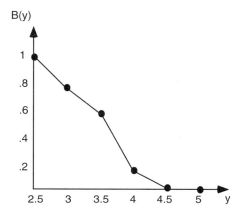

Figure 5.8   Fuzzy set expressing the
concept of the salary of young employees.

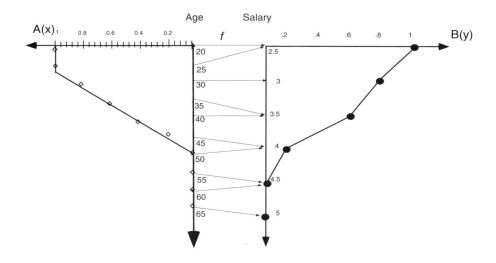

Figure 5.9   Illustration of the extension principle.

for any $A \in \mathcal{F}(X), y \in Y$, and

$$[\tilde{f}^{-1}(B)](x) = B(f(x)) \tag{5.4}$$

for any $B \in \mathcal{F}(Y)$ and $x \in X$.

When f is a continuous function defined on the set of real numbers, Eq. (5.3) must be replaced with the more general formulation

$$[\tilde{f}(A)](y) = \sup_{x|y=f(x)} A(x) \tag{5.5}$$

where sup denotes the supremum (see Chapter 6).

Given a function f that maps elements of a set X to elements of set Y, the extension principle provides us with rules by which f is extended either to function $\tilde{f}$, which maps fuzzy sets on X to fuzzy sets on Y, or to function $\tilde{f}^{-1}$, which maps fuzzy sets on Y to fuzzy sets on X. Given a fuzzy set A on X, function $\tilde{f}$ allows us to determine a fuzzy set on Y that is induced by the given function f. Conversely, given a fuzzy set B on Y, function $\tilde{f}^{-1}$ allows us to determine a fuzzy set on X induced by f. Function $\tilde{f}$ is defined by Eq. (5.3) if X and Y are finite sets or by Eq. (5.5) for arbitrary sets; function $\tilde{f}^{-1}$ is defined by Eq. (5.4). A visual summary of the various aspects of the extension principle is given in Fig. 5.10.

Let f and B denote, respectively, the function defined in Table 5.1 and a fuzzy set that attempts to capture, in a given context, the concept of a low salary. Assume that

$$B = 1/2.5 + 0.75/3 + 0.5/3.5 + 0.25/4 + 0/4.5 + 0/5$$

Then according to Eq. (5.4) we obtain the fuzzy set

$$\begin{aligned}
\tilde{f}^{-1}(B) = {}& B(f(20))/20 + B(f(25))/25 + B(f(30))/30 + B(f(35))/35 \\
& + B(f(40))/40 + B(f(45))/45 + B(f(50))/50 + B(f(55))/55 \\
& + B(f(60))/60 + B(f(65))/65 = 1/20 + 1/25 + 0.75/30 + 0.5/35 \\
& + 0.5/40 + 0.25/45 + 0.25/50 + 0/55 + 0/60 + 0/65
\end{aligned}$$

This fuzzy set is defined on X and represents the age of employees with low salaries.

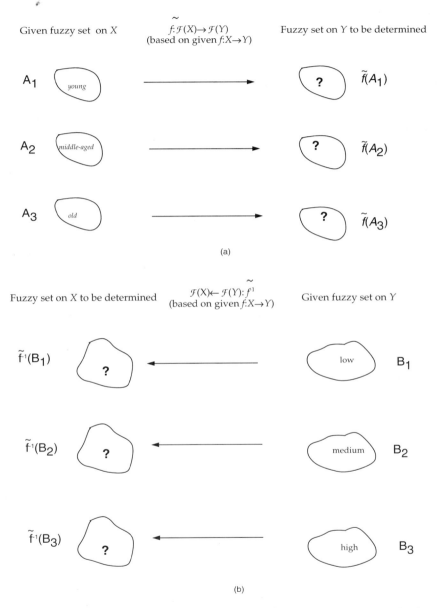

Figure 5.10   Extension principle: (a) Equation (5.3); (b) Equation (5.4).

5.5 MEASUREMENT OF FUZZINESS

As already mentioned, classical sets may be viewed as special fuzzy sets, called crisp sets, whose membership grades are restricted to 0 and 1 values. These sets are in some sense degenerate since they are not capable of expressing gradual transitions from membership to nonmembership; their boundaries are always sharp and, consequently, they are devoid of fuzziness. Any set that is not crisp involves some degree of fuzziness which results from the imprecision of its boundaries. The less precise the boundary, the more fuzzy the set. Different fuzzy sets are thus fuzzy to different degrees.

Let the degree to which a set is fuzzy be called its *fuzziness*. To measure fuzziness means to assign a nonnegative number to each fuzzy set. Clearly, these numbers cannot be arbitrary. They must satisfy some requirements that can easily be justified on intuitive grounds as essential for capturing the concept of fuzziness. To discuss these requirements, let $f(A)$ denote the measure of fuzziness of a fuzzy set A.

One obvious requirement is that the measure of fuzziness should be zero for all crisp sets and greater than zero for all other sets. That is, $f(A) \geq 0$ for all fuzzy sets and $f(A) = 0$ if and only if A is a crisp set.

Another requirement is based on our intuition that the sharper the boundary of a fuzzy set, the less fuzzy the set is. We can thus connect the fuzziness of a set to the sharpness of its boundary, which is easy to comprehend. In general, the sharpness of a boundary of a fuzzy set is determined by the closeness of its membership grades to the ideal values 0 and 1. The closer the membership grades are to the ideal values, the sharper the boundary. However, not all fuzzy sets are comparable in terms of the sharpness of their boundaries. To be comparable the membership grades of one fuzzy set must be closer or equally close to the ideal values 0 and 1 for all elements of the universal set. For any two fuzzy sets A and B that are comparable in this sense, if A has a sharper boundary than B, then A cannot be more fuzzy than B. That is, we must require that $f(A) \leq f(B)$ for these fuzzy sets.

Our intuition allows us also to identify the fuzzy set for which the highest fuzziness is obtained. It is the fuzzy set whose membership grades are as far from the ideal values as possible. That is, $f(A)$ is required to obtain its maximum when $A(x) = 0.5$ for all $x \in X$.

Various ways of measuring fuzziness have been proposed in the literature, all of which satisfy these three essential requirements. Each way of measuring is based on some particular view of fuzziness.

We describe only one, in which the fuzziness of any set is measured by the lack of distinction between the set and its complement. This makes sense since it is precisely the lack of distinction between sets and their complements that distinguishes fuzzy sets from crisp sets.

Given a fuzzy set A, the simplest way of expressing the local distinction between the membership grade $A(x)$ and its complement $\overline{A}(x) = 1 - A(x)$ for each $x \in X$ is the absolute value of their difference, that is,

$$|A(x) - [1 - A(x)]| = |2A(x) - 1|$$

Since the largest value of this difference is 1, the lack of local distinction is expressed by

$$1 - |2A(x) - 1|$$

When X is finite, fuzziness, $f(A)$, is then measured by the sum of all these local values:

$$f(A) = \sum_{x \in X} (1 - |2A(x) - 1|) \tag{5.6}$$

Notice that

$$0 \leq f(A) \leq |X|$$

the lower bound is obtained if and only if A is a crisp set while the upper bound is obtained only for the set for which $A(x) = 0.5$ for all $x \in X$.

As an example, let us apply Eq. (5.6) to fuzzy sets A and B defined by their graphs in Fig. 5.9. We obtain

$$A(x) = 10 - 1 - 1 - 0.6 - 0.2 - 0.2 - 0.6 - 1 - 1 - 1 - 1 = 2.4$$
$$B(x) = 6 - 1 - 0.6 - 0.2 - 0.6 - 1 - 1 = 1.6$$

When X is an interval of real numbers, Eq. (5.6) can be readily modified for calculating $f(A)$ by replacing the summation with integration. This should be explored by students who have a sufficient background in the calculus.

EXERCISES

5.1 What is the relationship between a fuzzy set and its α-cuts? What is the significance of this relationship?

5.2 Let A be a fuzzy set defined on the universal set $\{a, b, c, d, e\}$, whose membership function is given by
$$A = 0.1/a + 0.5/b + 1/c + 1/d + 0.7/e$$
    (a) Find all different α-cuts of A.
    (b) Determine the standard fuzzy complement $\overline{A}$ and all its different α-cuts.
    (c) What are the supports, heights, and cores of A and $\overline{A}$?
    (d) Determine all different α-cuts of $A \cap \overline{A}$ and $A \cup \overline{A}$.

5.3 Let fuzzy set B be the standard complement of the fuzzy set A in Exercise 5.2. Find the support, height, and core of $A \cap B$ and $A \cup B$, where the standard fuzzy set operations are used for the intersection and the union.

5.4 Let A be a fuzzy set of a universal set X. Show that the height of the fuzzy set $A \cap \overline{A}$ is not greater than 0.5. Also show that the height of the fuzzy set $A \cup \overline{A}$ is always greater than or equal to 0.5.

5.5 Is the standard fuzzy complement cutworthy? Justify your answer.

5.6 Show how membership function of fuzzy set A and its complement $\overline{A}$ in Exercise 5.2 can be reconstructed from their α-cuts via formula (5.2).

5.7 Determine analytic expressions for α-cuts of fuzzy sets A and B in Exercise 4.6.

5.8 Using the generic forms of triangular-shaped and trapezoidal-shaped membership functions introduced in Sec. 4.3, determine generic forms of their α-cuts.

5.9 Consider Table 5.1 and the associated example discussed in Sec. 5.4. Using the extension principle, determine the fuzzy set B characterizing salaries of old employees provided that the meaning of the concept *old* is expressed by the fuzzy set
$$A = 0/20 + 0/25 + 0/30 + 0/35 + 0.2/40 + 0.4/45 + 0.6/50 + 0.8/55 + 1/60 + 1/65$$

5.10 It is well established that the maximum heart rate is a function of age, which is given in a discrete form in Table 5.2. Define, within the given age range, reasonable fuzzy sets to capture the concepts of *young, middle age,* and *old.* Determine then, using the extension principle, the corresponding fuzzy sets that characterize the maximum heart rate of young, middle-aged, and old people.

5.11 Prove Theorem 5.1.

5.12 Prove Theorem 5.2.

TABLE 5.2 MAXIMUM HEART RATE AS A FUNCTION OF AGE

Age	20	22	24	26	28	30	32	34
Maximum heart rate	200	198	196	194	192	190	189	187

Age	36	38	40	45	50	55	60	65
Maximum heart rate	186	184	182	179	175	171	160	150

6

CLASSICAL RELATIONS

6.1 INTRODUCTION

A classical (or crisp) relation represents the presence or absence of association, interaction, or connection between elements of two or more sets taken in a certain order. For example, if we are told that two people, Jim and Carol, are related, we understand that there is some familial connection between them. Thus the pair ⟨Jim, Carol⟩ is said to satisfy some relationship, say *brother of.* This is a relationship that holds between them, but not between some other pairs. Further, we can see in this example that the relation *brother of* imposes an order on the two members of the relation. The relation arranges them in a left-to-right way: Jim is the brother of Carol, but not the other way around. Generalizing this simple example, we may distinguish certain ordered pairs from any other ordered pairs because the distinguished pairs satisfy some given relation and the others do not. We confine our discussion of relations first to relations involving just two sets of elements. These relations are called *binary relations.*

Familial relations are obvious examples of relations, but there are many others commonly used. Two of the most familiar relations are ordering relations, *greater than* and *less than,* which are applied in may different contexts. In all of these cases, we see that two individuals, objects, or entities are paired and that each pair exhibits—or satisfies—the relation in question. Of course there are many possible pairs of things in the world. However, only some of them satisfy a given relation so it becomes important to understand the full nature of these possible pairs. For this, we first review the concept of a Cartesian product and then employ it for defining formally the concept of a relation.

As introduced in Sec. 3.5, the Cartesian product, $A \times B$, of two sets, A and B (in this order), is the set of all possible ordered pairs constructed by combining elements of A with elements of B such that the first element in each pair is a member of A and the second element is a member of B. That is, this product is the set of all ordered pairs $\langle a, b \rangle$, where $a \in A$ and $b \in B$.

In the ordered pair composed of elements a and b—represented as $\langle a, b \rangle$—it matters in which order the elements appear. That is, $\langle a, b \rangle$ is a different pair from $\langle b, a \rangle$.

As an example, consider two variables x and y whose possible values are real numbers in the closed intervals $[x_1, x_2]$ and $[y_1, y_2]$, respectively. Then the Cartesian product $[x_1, x_2] \times [y_1, y_2]$ consists of all points in the rectangle shown in Fig. 6.1. Consider now sets $A = \{1, 2\}$ and $B = \{a, b, c\}$. Then,

$$A \times B = \{\langle 1, a \rangle, \langle 1, b \rangle, \langle 1, c \rangle, \langle 2, a \rangle, \langle 2, b \rangle, \langle 2, c \rangle\}$$

$$B \times A = \{\langle a, 1 \rangle, \langle b, 1 \rangle, \langle c, 1 \rangle, \langle a, 2 \rangle, \langle b, 2 \rangle, \langle c, 2 \rangle\}$$

$$A \times A = \{\langle 1, 1 \rangle, \langle 1, 2 \rangle, \langle 2, 1 \rangle, \langle 2, 2 \rangle\}$$

$$B \times B = \{\langle a, a \rangle, \langle a, b \rangle, \langle a, c \rangle, \langle b, a \rangle, \langle b, b \rangle, \langle b, c \rangle, \langle c, a \rangle,$$
$$\langle c, b \rangle, \langle c, c \rangle\}$$

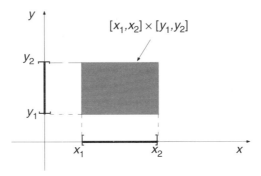

Figure 6.1 Geometric representation of Cartesian product.

By taking a subset of a given Cartesian product $X \times Y$, we define a relation R between elements of set X and those of set Y. That is,

$$R \subseteq X \times Y$$

Pairs of elements that are included in R are viewed as related; the remaining pairs are not. Since two sets are involved in this relation, it is called a binary relation.

As an example of a binary relation, consider two variables x and y whose values are real numbers in the intervals $[x_1, x_2]$ and $[y_1, y_2]$, respectively. Assume that the variables always have equal values. This equality relation, which we denote by E, is expressed by the equation

$$x = y, \text{ where } x \in [x_1, x_2] \text{ and } y \in [y_1, y_2]$$

and illustrated in Fig. 6.2. That is, $E \subset [x_1, x_2] \times [y_1, y_2]$.

Relations may also be defined on Cartesian products of three or more sets. We refer to these relations as *ternary* (with three sets), *quaternary* (with four sets), *quinary* (with five sets), and so on. In general, a relation defined on the Cartesian product of n sets is called an n-dimensional relation ($n \geq 2$). Thus,

$$R \subseteq X_1 \times X_2 \times \ldots \times X_n$$

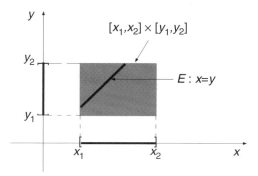

Figure 6.2   Equality relation E between variables x and y.

is an n-dimensional relation. The Cartesian product $X_1 \times X_2 \times \dots \times X_n$ is in this case the set of all n-tuples constructed by combining elements of each set with elements of the remaining $n - 1$ sets such that the first element in each n-tuple is a member of X_1, the second element is a member of X_2, and so forth.

The sets involved in a Cartesian product may be equal. In such a case, the symbol X^n is often used as shorthand notation instead of

$$\underbrace{X \times X \times \dots \times X}_{n - \text{times}}$$

for any $n \geq 2$.

Consider, for example, the set of all points in a solid body in the three-dimensional Euclidean space, such as the one shown in Fig. 6.3. This set is clearly a subset of

$$\mathbb{R}^3 = \mathbb{R} \times \mathbb{R} \times \mathbb{R}$$

and, hence, it is a ternary relation in the three-dimensional Euclidean space.

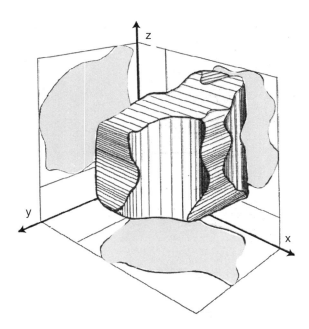

Figure 6.3　Ternary relation in three-dimensional Euclidean space.

As another example of a ternary relation, consider a regular long-distance flight which is always operated by three copilots. According to one operation rule, it is required that at least two of the copilots be on duty at any time during the flight; that is, only one of them may sleep or otherwise rest. Let $A_i = \{d_i, r_i\}$, where d_i, r_i denote that copilot i is *on duty* or *rests* respectively (i = 1, 2, 3). Then, the operation rule may be expressed as the ternary relation

$$R = \{\langle d_1, d_2, d_3 \rangle, \langle d_1, d_2, r_3 \rangle, \langle d_1, r_2, d_3 \rangle, \langle r_1, d_2, d_3 \rangle\}$$

defined on $A_1 \times A_2 \times A_3$. From among eight ordered triples of this Cartesian product, only four triples satisfy this relation.

Although we discuss some general properties of n-dimensional relations in Sec. 6.6, most attention is given in this text to binary relations, especially those defined on the Cartesian product of a single set. Among the various types of relations, we examine equivalence relations and ordering relations more closely.

6.2   REPRESENTATIONS

As shown in the previous section, every relation on a finite Cartesian product can be defined by listing all pairs (or, more generally, n-tuples) that belong to the relation. However, there are various visual representations of binary relations which are more appealing to our intuition. We introduce some of them in this section.

Coordinate Representation

We encountered a graphical representation of relations in Fig. 6.1. We can also generate a slightly different form of this two-dimensional, coordinate representation. Consider an agricultural example illustrating the relation *is the product of*:

$$R = \{\langle \text{eggs, hens} \rangle, \langle \text{milk, cows} \rangle, \langle \text{milk, goats} \rangle\}$$

The coordinate diagram of this relation is shown in Fig. 6.4, where the dots specify which pairs are included in the relation. Note that the additional entity *corn* is not matched with any of the entities of the x-coordinate. Hence, we can see that it is not a participant in the relation R.

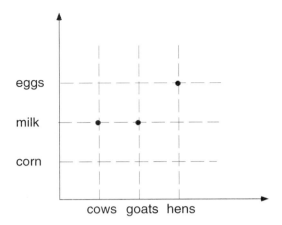

Figure 6.4   Coordinate diagram of a relation.

Matrix Representation

In matrix representation, we follow coordinate representation, but at each intersection representing the relation we place a 0 if the relation does not exist between the two individuals and a 1 if it does. This is because we may describe the characteristic function for a relation in terms of assigning these two labels:

$$\chi_R(x, y) = \begin{cases} 1 \text{ if } \langle x, y \rangle \in R \\ 0 \text{ otherwise} \end{cases}$$

Accordingly, the relation *is the product of* is represented in Fig. 6.5. Again, corn is not a participant in the relation R and, hence, is assigned a 0 in all entries of its row.

The significance of this form of representation is that it—like many of our previous representations of crisp concepts—relies on the two-valued, Yes/No device of depicting the fact that a crisp relation either definitely holds between two entities or definitely does not: It holds with degree 1 or degree 0. Moreover, this representation is particularly suitable for dealing with binary relations on a computer. For n-dimensional relations, matrices are replaced with n-dimensional arrays.

R	cows	goats	hens
eggs	0	0	1
milk	1	1	0
corn	0	0	0

Figure 6.5 Matrix representation of a relation.

Mappings

In another visual representation of relations, members of one set are being linked with certain members of another set. We say that members of the first set are *mapped to* members of the second set. Fig. 6.6 shows the visual representation of the relation *is the product of* in which the first set is {eggs, milk, corn} and the second set is {cows, goats, hens}. There are two aspects of this representation worthy of note. First, the element *corn* is not mapped to anything in the second set; this replicates the assignment of 0 in all columns of the row labeled corn in the matrix representation. Second, the element *milk* is mapped to two elements in the second set. Thus, given a relation, an element in the first set may be mapped to more than one element of the second set.

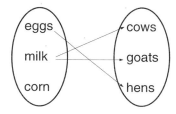

Figure 6.6 Mapping representation of a relation.

Binary relations in which no element of the first set is mapped to more than one element of the second set are called *functions*. No restriction is made on the number of elements of the first set that are mapped to the same element of the second set for functions, however.

A special notation is usually used for functions. Given a binary relation

$$R \subseteq A \times B$$

which is a function, it is usual to express it in the form

$$R : A \to B$$

An example of a function is given in Fig. 6.7.

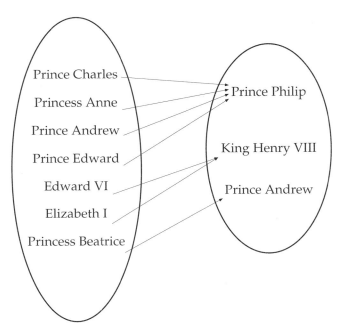

Figure 6.7   The function *is the biological ancestor of.*

Directed Graph

For any binary relation that is defined on the Cartesian product of a single set, a representation by a *directed graph* is particularly illuminating, even though all the other representations of binary relations are applicable as well. The directed graph representing a relation $R \subseteq X^2$ can be conveniently expressed by a diagram with the following properties: (i) each element of the set X is represented by a node (a small circle) in the diagram; and (ii) directed connections between nodes indicate pairs of elements that are included in the relation (i.e., elements that are related).

As an example, consider the set C of six courses labeled c_1, c_2, ..., c_6 whose prerequisite requirements are described by the diagram in Fig. 6.8a. This diagram represents a binary relation on C^2 according to which course c_i is related to course c_j $(i, j = 1, 2,..., 6)$ if c_i is a prerequisite for c_j. For example, c_1 is a prerequisite for all the remaining courses, c_3 is a prerequisite for c_5 and c_6, c_4 is not a prerequisite for any of the courses, and so forth. We may also define an associated relation, which is simpler but more practical, that characterizes only the immediate prerequisites. A diagram of this relation is shown in Fig. 6.8b.

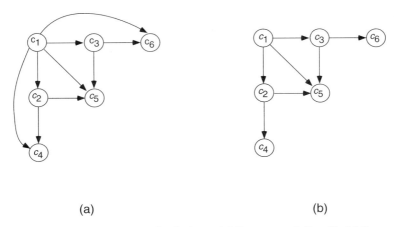

(a)   (b)

Figure 6.8   Diagrams of relations: (a) "a prerequisite of"; (b) "an immediate prerequisite of."

6.3 OPERATIONS ON BINARY RELATIONS

Since relations are basically sets, all concepts defined for sets are applicable to them as well. This means, for example, that all operations on sets (complement, union, intersection) are also applicable to relations. However, some additional operations, which have no counterparts for sets, are also applicable to relations. In this section, we restrict discussion to two basic operations on binary relations. Some operations that are applicable to any n-dimensional relation are discussed in Sec. 6.6.

Given a binary relation R, its *inverse* R^{-1} is defined by exchanging the order of elements in all pairs contained in the relation. Hence, if

$$R \subseteq X \times Y$$

then

$$R^{-1} \subseteq Y \times X$$

Clearly, the inverse of R^{-1} is the original relation R. That is,

$$(R^{-1})^{-1} = R$$

for any binary relation R.

Observe, for example, that the inverse of the relation in Fig. 6.6 is a function though the given relation is not a function. On the other hand, the relation in Fig. 6.7 is a function while its inverse is not a function.

An important operation on binary relations is *composition*. By this operation, two compatible binary relations are combined and one binary relation is obtained as the result. To be compatible for composition, relations P and Q must be defined as

$$P \subseteq X \times Y$$
$$Q \subseteq Y \times Z$$

That is, the second set in P must be the same as the first set in Q. The composition of these compatible relations P and Q, denoted by $P \circ Q$, is defined as the relation

$$R \subseteq X \times Z$$

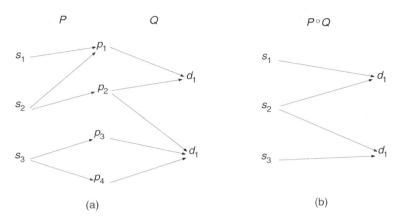

Figure 6.9   Example of the composition of two binary relations.

such that $\langle x, z \rangle \in R$ if and only if there exists at least one element y in set Y for which the pair $\langle x, y \rangle$ is in P and the pair $\langle y, z \rangle$ is in Q.

As an example of the use of the operation of composition, let us consider the following compatible relations: P is a relation between a set of three medical symptoms, s_1, s_2, s_3, and a set of four patients, p_1, p_2, p_3, p_4; Q is a relation between the same set of patients and a set of two diseases, d_1 and d_2. Assuming the relations given in Fig. 6.9a, their composition is the relation between symptoms and diseases shown in Fig. 6.9b. Knowing the symptoms manifested by individual patients and knowing also the diseases of these patients, the composition allows us to determine possible connections between symptoms and diseases.

It is easy to verify that the composition is an associative operation. That is,

$$(P \circ Q) \circ R = P \circ (Q \circ R)$$

provided that the relations P, Q and Q, R are compatible. The composition is not commutative. That is, in general,

$$P \circ Q \neq Q \circ P$$

However, it always satisfies the property

$$(P \circ Q)^{-1} = Q^{-1} \circ P^{-1}$$

6.4 EQUIVALENCE AND COMPATIBILITY RELATIONS

Among the great variety of possible relations on a single set X (i.e., subsets of X^2), three important types of relations are introduced and examined in this book. These are equivalence relations, compatibility relations, and ordering relations. In this section, we cover equivalence and compatibility relations. Their fuzzy counterparts are discussed in Sec. 7.4.

In each equivalence relation on a given set X, elements of X are related if they are equivalent in terms of a specified characteristic. To illustrate this concept, let us use a simple example.

Consider the set of ten students listed in Table 6.1 who took the same course. Assume that four characteristics of each student in this set are of interest: his or her grade in the course, major field of study, age, and the status as either a full-time or part-time student. Choosing one of these characteristics, or a combination of them, we may define a particular equivalence relation. We consider students who are not distinguished by the chosen characteristic as equivalent in terms of this characteristic. Thus, for example, Bob is equivalent to Cliff and Debby is equivalent to George in terms of their grades, but Alan is not equivalent to Bob in this sense. We say that students who are equivalent according their grades in the course are related by the equivalence relation *having an equal grade.*

TABLE 6.1   ILLUSTRATION OF EQUIVALENCE RELATIONS

Student	Grade	Major	Age	Full-time/ part-time
Alan	B	Biology	19	Full-time
Bob	C	Physics	19	Full-time
Cliff	C	Mathematics	20	Part-time
Debby	A	Mathematics	19	Full-time
George	A	Mathematics	19	Full-time
Jane	A	Business	21	Part-time
Lisa	B	Chemistry	21	Part-time
Mary	C	Biology	19	Full-time
Nancy	B	Biology	19	Full-time
Paul	B	Business	21	Part-time

Since the names of students have distinct first letters in our example, we can abbreviate the names by their first letters. Then our set of students, X, is defined as

$$X = \{A, B, C, D, G, J, L, M, N, P\}$$

To define our equivalence relation *having an equal grade* on this set, we have to decide for each ordered pair in the Cartesian product X^2 whether the criterion of equal grades is satisfied or not. If it is satisfied, the pair is included in our equivalence relation; otherwise, it is not included.

It is obvious that all the pairs $\langle A, A \rangle, \langle B, B \rangle, \langle C, C \rangle, \ldots$ must be included in the equivalence relation since each student is equivalent to himself/herself in terms of the grade (as well as in any other characteristic). This property of equivalence relations is called *reflexivity*.

Another property of equivalence relations is *symmetry* in terms of the ordered pairs included in the relation. That is, if an ordered pair is included in the relation, then the pair with the inverse order of elements must also be included in it. For example, since Bob is equivalent to Cliff in terms of grades, it is also the case that Cliff is equivalent to Bob in this regard. Hence, both pairs $\langle B, C \rangle$ and $\langle C, B \rangle$ must be included in this equivalence relation.

The third distinguishing property of equivalence relations is *transitivity*. According to this property, if one element is equivalent to another element and the latter is equivalent in turn to a third element, then the first element is also equivalent to the third element. For example, since Alan is equivalent to Lisa and Lisa is equivalent to Mary in terms of their grades, Alan is also equivalent to Mary in this respect. Hence, the inclusion of pairs $\langle A, L \rangle$ and $\langle L, M \rangle$ in this equivalence relation implies that the pair $\langle A, M \rangle$ must also be included in it.

These three properties—reflexivity, symmetry, and transitivity—which capture our intuition about the notion of equivalence, form a mathematical basis for formulating the concept of an equivalence relation in precise, mathematical terms. To define these properties formally, let R denote a binary relation on X. Then

(a) R is *reflexive* if and only if $\langle x, x \rangle \in R$ for each $x \in R$;

(b) R is *symmetric* if and only if for every pair of elements in X, either both pairs $\langle x, y \rangle$ and $\langle y, x \rangle$ are included in R or neither of them is included in R;

(c) R is *transitive* if and only if for any three elements x, y, z in X, $\langle x, z \rangle \in R$ whenever both $\langle x, y \rangle \in R$ and $\langle y, z \rangle \in R$.

Any binary relation on a single set that satisfies these three properties is called an *equivalence relation*. Such a relation clearly expresses our intuitive understanding of equivalence among individuals of any kind in terms of some specified characteristic of these individuals.

The relation expressing equivalence of students listed in Table 6.1 with respect to their grades is expressed by the matrix in Table 6.2. As explained earlier, entries of this matrix are values of the characteristic function of this equivalence relation: 1 indicates that students in the associated pair are equivalent; 0 means that they are not equivalent.

TABLE 6.2 EQUIVALENCE RELATION R DEFINED ON THE SET OF STUDENTS LISTED IN TABLE 6.1 WITH RESPECT TO THEIR GRADES.

R	A	B	C	D	G	J	L	M	N	P
A	1	0	0	0	0	0	1	0	1	1
B	0	1	1	0	0	0	0	1	0	0
C	0	1	1	0	0	0	0	1	0	0
D	0	0	0	1	1	1	0	0	0	0
G	0	0	0	1	1	1	0	0	0	0
J	0	0	0	1	1	1	0	0	0	0
L	1	0	0	0	0	0	1	0	1	1
M	0	1	1	0	0	0	0	1	0	0
N	1	0	0	0	0	0	1	0	1	1
P	1	0	0	0	0	0	1	0	1	1

The same equivalence relation is also expressed by the simplified diagram in Fig. 6.10a. In this diagram, the loop from each node to itself (as required by reflexivity) is omitted and connections between nodes are not directed since the property of symmetry guarantees that all existing connections appear in both directions.

A diagram of another equivalence relation defined on the set of students, one based on their major fields of study, is shown in Fig. 6.10b.

We can see that both equivalence relations expressed in Fig. 6.10 partition the set of students into subsets such that students in

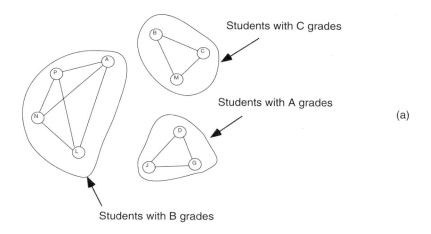

Students with C grades

Students with A grades

(a)

Students with B grades

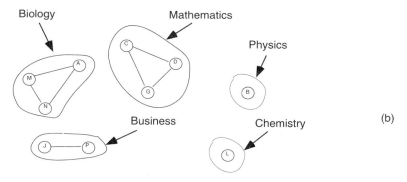

Biology

Mathematics

Physics

Business   Chemistry   (b)

Figure 6.10   Equivalence relations defined on the set of students listed in Table 6.1.

each subset are related to each other while they are not related to students in any of the other subsets. These subsets are called *equivalence classes*. Each of them consists of students who are equivalent in terms of the relevant characteristic. Thus, for example, students with B grades form an equivalence class of the first relation, while students majoring in mathematics form an equivalence class of the second relation.

It is easy to see that the partitioning of a given set X into equivalence classes by a given equivalence relation on X is a general property of all equivalence relations.

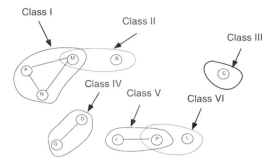

Figure 6.11   Compatibility relation defined on the set of students listed in Table 6.1.

Generalizations of equivalence relations are relations that are required to be reflexive and symmetric, but not necessarily transitive. These relations are called *compatibility relations*. Elements that are related by a compatibility relation are viewed as compatible (in some prescribed way), but not necessarily equivalent.

As an example, consider two students listed in Table 6.1 as compatible if they do not differ in more than one of the four characteristics. This notion of compatibility induces a compatibility relation whose diagram is shown in Fig. 6.11. We can see that the set of students is classified into subsets of compatible students. These are called *compatibility classes* and those of them that are not contained with other compatibility classes are called *maximal compatibility classes*. In our example, there are six maximal compatibility classes, labeled in Fig. 6.11 by I through VI. For example, Alan, Mary, and Nancy are compatible students according to our conception of compatibility, and Mary (but not Alan and Nancy) is also compatible with Bob. We can also see that Cliff is not compatible with any student except himself and, hence, he forms his own maximal compatibility class.

6.5 PARTIAL ORDERINGS

Ordering relations of various types are very important binary relations on a single set. As the name suggests, the purpose of ordering relations is to describe various ways in which individuals of a given set may be usefully ordered. There are many examples of useful orderings: ordering of courses in a program by prerequisite requirements; ordering of people by their age, height, weight, salary, and so forth; ordering of countries by the size of their populations, gross national income, or by other criteria.

To capture the intuitive notion of ordering, notice that any ordering relation must be transitive and must not be symmetric. Indeed, if x precedes y according to some ordering, then y cannot precede x by the same ordering as well. Additionally, if x precedes y and y precedes z, then x precedes z as well.

The most fundamental type of ordering relations are relations that are called *partial orderings*. They play an important role in the foundations of mathematics as well as in numerous applications.

In addition to being transitive, partial orderings are relations that are also reflexive and asymmetric in a strong sense, which is referred to as *antisymmetry*. A relation R on X is antisymmetric if and only if for any x and y in X, if $\langle x, y \rangle \in R$ and $\langle y, x \rangle \in R$ then $x = y$.

The symbol $x \leq y$ is often used for describing a partial ordering of elements x, y of a given set. When we apply it to numbers, it means that x is less than or equal to y. When applied to a partial ordering R on an arbitrary set X, $x \leq y$ means that x precedes y according to R. That is, if $\langle x, y \rangle \in R$, we write $x \leq y$. If x precedes y, we also say that y *succeeds* x. When $x \leq y$, according to a partial ordering, x is called a *predecessor* of y, while y is called a *successor* of x. If $x \leq y$ and there is no $z \in X$ such that $x \leq z$ and $z \leq y$, then x is called an *immediate predecessor* of y, and y is called an *immediate successor* of x.

As an example of partial ordering, let us consider the power set $\mathcal{P}(X)$ of an arbitrary, nonempty set X. Elements of $\mathcal{P}(X)$ (i.e., subsets of X) can be ordered by the relation of set inclusion $\subseteq$. For any ordered pair $\langle A, B \rangle$ of sets A, B in $\mathcal{P}(X)$, we say that A is smaller than or equal to B (and we write $A \leq B$) if and only if $A \subseteq B$. To show that this ordering is partial, we have to demonstrate that it is reflexive, transitive, and antisymmetric. This can easily be accomplished: $\leq$ is reflexive since $A \subseteq A$ for any set; $\leq$ is transitive since $A \subseteq B$ and $B \subseteq C$ implies $A \subseteq C$; $\leq$ is antisymmetric since $A \subseteq B$ and $B \subseteq A$ implies $A = B$.

The partial ordering of subsets of the set $\{x_1, x_2, x_3\}$ is captured by the diagram in Fig. 6.12. In this diagram, each subset is connected only to its immediate predecessors and immediate successors. The connections are directed in order to distinguish predecessors from successors: the arrow $\leftarrow$ indicates the inequality $\leq$. Diagrams of this sort, which are used for describing partial orderings, are called *Hasse diagrams*.

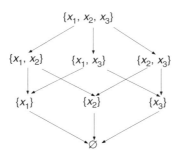

Figure 6.12 Partial ordering of subsets of a set with three elements.

Any set of real numbers can be partially ordered by the usual inequality relation $\leq$. As an example, the Hasse diagram of the partial ordering of decimal digits 0, 1, ..., 9 is shown in Fig. 6.13. This ordering, which is a special case of partial ordering, is called a *linear ordering*. In addition to reflexivity, transitivity, and antisymmetry, each linear ordering on set X requires that either $a \leq b$ or $b \leq a$ for all $a, b \in X$. Let us use the linear ordering of real numbers to introduce a few concepts pertaining to real numbers that are needed for proper understanding of some properties of fuzzy sets.

Let X be a subset of the set $\mathbb{R}$ of real numbers. If there is a real number u such that $x \leq u$ for every x in X, then u is called an *upper bound* of X, and we say that X is *bounded above*. If, in addition, no number smaller than u is an upper bound of X, then u is called the *least upper bound* or *supremum* of X, and we write

$$u = \sup X$$

When sup X is a member of X, then it is the *maximum number* in X, and we write

$$u = \max X$$

Figure 6.13 Linear ordering of decimal digits, a special case of partial ordering.

Consider now a real number l such that $l \leq x$ for every x in X. Then l is called a *lower bound*, and we say that X is *bounded below*. If, in addition, no number greater than l is a lower bound, then l is called the *greatest lower bound* or *infimum* of X, and we write

$$l = \inf X$$

When $\inf X$ is a member of X, then it is the *minimum number* in X, and we write

$$l = \min X$$

As an example, let $X = [0,1]$. Then, X is bounded both above and below, and

$$\sup X = \max X = 1$$
$$\inf X = \min X = 0$$

Consider now the set $X = (0, 1)$. Then, X is again bounded from both above and below, and

$$\sup X = 1$$
$$\inf X = 0$$

However, there is no minimum number and no maximum number in this set.

Consider now the Cartesian product of n given sets. Its elements (ordered n-tuples) can be partially ordered in the following way: given any pair

$$\langle a_1, a_2, ..., a_n \rangle \text{ and } \langle b_1, b_2, ..., b_n \rangle$$

of the Cartesian product, we define

$$\langle a_1, a_2, ..., a_n \rangle \leq \langle b_1, b_2, ..., b_n \rangle$$

if and only if

$$a_1 \leq b_1, a_2 \leq b_2, ..., a_n \leq b_n$$

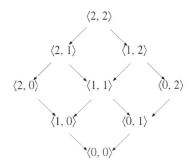

Figure 6.14   Partial ordering of elements
of the Cartesian product {0,1,2}×{0,1,2}.

It is easy to verify that this ordering is reflexive, transitive, and antisymmetric. Hence, it is a partial ordering. The Hasse diagram in Fig. 6.14 exemplifies this partial ordering by using the Cartesian product

$$\{0, 1, 2\} \times \{0, 1, 2\}$$

Other examples of partial orderings are refinement orderings of partitions of given sets. An example of these refinement orderings is shown for partitions of the set {1, 2, 3} in terms of the Hasse diagram in Fig. 6.15.

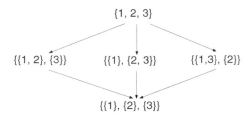

Figure 6.15   Refinement ordering of partitions
of the set {1,2,3}.

6.6 PROJECTIONS AND CYLINDRIC EXTENSIONS

Projections and cylindric extensions are operations that are applicable to any n-dimensional relations ($n \geq 2$). These two operations are in some sense inverses of each other. Their names are borrowed from geometry.

A projection of an n-dimensional relation R into a specified subset of k dimensions, where $1 \leq k \leq n$, is an operation that produces a k-dimensional relation that is in some sense compatible with R. This k-dimensional relation consists of all k-tuples that are obtained from the n-tuples in R by ignoring the components that correspond to the remaining n-k dimensions.

As an example, consider the solid object depicted in Fig. 6.16. The set of all points in this object may be viewed as a relation in the three-dimensional Euclidean space. Also shown in the figure are the usual geometric projections of the object into three planes, each represented by two of the three coordinates. These geometric projections are exactly the same as the projections of the three-dimensional relation into the three pairs of dimensions. Each of the two-dimensional projections may further be projected into either of their respective dimensions. For example, one-dimensional projections of a two-dimensional relation R (a rectangle) are shown in Fig. 6.17. These projections are intervals of real numbers R_x and R_y.

As another example, consider the three-dimensional relation R on $\{0, 1\}^3$ that consists of the four triples

$$
\begin{array}{llll}
R: & 0 & 0 & 1 \\
& 0 & 1 & 1 \\
& 1 & 1 & 0 \\
& 1 & 1 & 1
\end{array}
$$

Denoting the projections to two of the three dimensions by R_{12}, R_{13}, R_{23}, we have:

$$
\begin{array}{lll}
R_{12}: & 0 \quad 0 \\
& 0 \quad 1 \\
& 1 \quad 1
\end{array}
\qquad
\begin{array}{lll}
R_{13}: & 0 \quad 1 \\
& 1 \quad 0 \\
& 1 \quad 1
\end{array}
\qquad
\begin{array}{lll}
R_{23}: & 0 \quad 1 \\
& 1 \quad 1 \\
& 1 \quad 0
\end{array}
$$

Given an n-dimensional relation R, its *cylindric extension* is an operation that produces the largest $(n + k)$-dimensional relation for

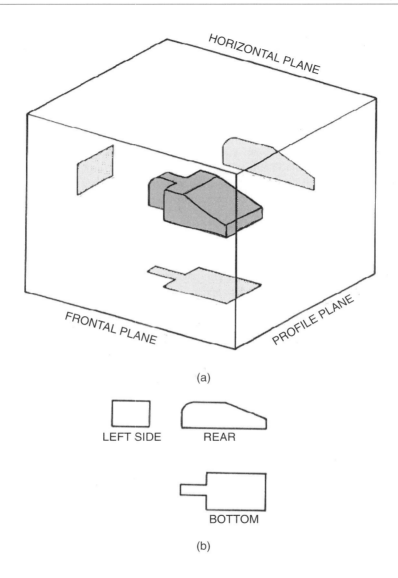

(a)

LEFT SIDE   REAR

BOTTOM

(b)

Figure 6.16   Projections of a three-dimensional object.

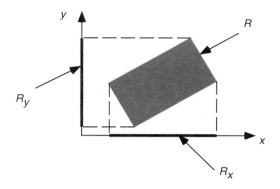

Figure 6.17   One-dimensional projections of a
relation in the two-dimensional Euclidean space.

some $k \geq 1$ that is compatible with R in the sense that its projection
into the original n dimensions is exactly the given relation R. The
cylindric extension of R consists of all $(n + k)$-tuples that can be
obtained by replicating each n-tuple of R and attaching to each replica
a distinct combination of elements in the added k dimensions. These
attached elements must be properly placed according to the specified
order of the dimensions. We use the symbol $\text{cyl}_d (R)$ to denote the
cylindric extension of relation R with respect to dimensions specified
by d (i.e., d is a list of dimensions).

For example, applying the operation of cylindric extension to
each of the three projections in the previous example with respect to
the missing dimension, we obtain

$\text{cyl}_3(R_{12})$:	0	0	**0**		$\text{cyl}_2(R_{13})$:	0	**0**	1		$\text{cyl}_1(R_{23})$:	**0**	0	1
	0	0	**1**			0	**1**	1			**1**	0	1
	0	1	**0**			1	**0**	0			**0**	1	1
	0	1	**1**			1	**1**	0			**1**	1	1
	1	1	**0**			1	**0**	1			**0**	1	0
	1	1	**1**			1	**1**	1			**1**	1	0

In each case, the bold entries indicate the dimension into which the cylindric extension is made.

It is known that the intersection of cylindric extensions of any projections of a given relation always contains the relation. However, they are equal only in special cases. In such cases, the relation is fully represented by the projections and we say that it is a *cylindric closure* of the given projections.

It can be verified that, in our previous example,

$$\text{cyl}_3(R_{12}) \cap \text{cyl}_2(R_{13}) \cap \text{cyl}_1(R_{23}) = R$$

That is, the three-dimensional relation R is fully represented by its three two-dimensional projections. Given the projections, the relation can be reconstructed by taking the intersection of their cylindric extensions. The relation is thus a cylindric closure of its two-dimensional projections.

The operations of projections and cylindric extensions involving three-dimensional Euclidean space are illustrated in Fig. 6.18.

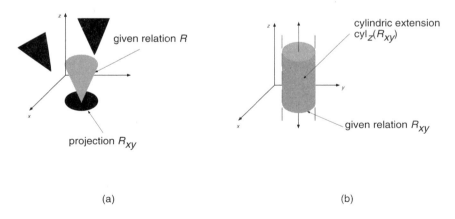

(a)   (b)

Figure 6.18   The operations of (a) projection and (b) cylindric extension.

EXERCISES

6.1 Find some examples of binary relations in your daily life. For each relation, find if it is reflexive, symmetric, antisymmetric, or transitive.

6.2 Given a group of people who know each other, define the relation characterized by the linguistic expression "a friend of." Which properties (symmetry, transitivity, etc.) does this relation possess?

6.3 Determine the matrix and diagram representations of the equivalence relations on the set of students given in Table 6.1 based on
   (a)   Their age,
   (b)   Their status as full-time or part-time students,
   (c)   Their major field of study and grade,
   (d)   Other combinations of the four characteristics in Table 6.1.

6.4 Order all partitions resulting from the equivalence relations obtained in Exercise 6.3 and those in Fig. 6.10 by the relation of partition refinement. Show that the relation is a partial ordering. Draw a simplified diagram of this partial ordering in which connections are shown only for immediate predecessors/successors.

6.5 Determine the matrix and diagram representations of the compatibility relations on the set of students given in Table 6.1 that are based on the following definitions of compatibility.
   (a)   Two students are compatible if they do not differ in more than two of the four characteristics.
   (b)   Two students are compatible if they do not differ in more than three of the four characteristics.

6.6 Let $A = \{1, 2, 3\}$. Which of the following binary relations on $A \times A$ is reflexive, symmetric, transitive, or antisymmetric?
   (a) $R_1 = \{\langle x, y \rangle \mid x < y\}$
   (b) $R_2 = \{\langle x, y \rangle \mid (x \leq y)\}$
   (c) $R_3 = \{\langle x, y \rangle \mid x = y\}$
   (d) $R_4 = \{\langle x, y \rangle \mid 2x = y\}$
   (e) $R_5 = \{\langle x, y \rangle \mid x = y - 1\}$
   (f) $R_6 = \{\langle 0, 0 \rangle, \langle 0, 1 \rangle, \langle 1, 0 \rangle, \langle 1, 1 \rangle, \langle 1, 2 \rangle, \langle 2, 2 \rangle, \langle 0, 2 \rangle, \langle 3, 3 \rangle\}$
   (g) $R_7 = \{\langle x, y \rangle \mid x < y \text{ or } x > y\}$
   (h) $R_8 = \{\langle 0, 0 \rangle, \langle 0, 3 \rangle, \langle 1, 1 \rangle, \langle 2, 2 \rangle, \langle 1, 0 \rangle, \langle 0, 1 \rangle, \langle 3, 1 \rangle, \langle 3, 3 \rangle, \langle 3, 0 \rangle, \langle 1, 3 \rangle$
   (i) $R_9 = \{\langle x, y \rangle \mid x = y \text{ or } x = y - 1 \text{ or } x - 1 = y\}$
   (j) $R_{10} = \{\langle 0, 1 \rangle, \langle 1, 3 \rangle, \langle 2, 1 \rangle, \langle 3, 2 \rangle\}$

6.7 Determine the inverse relation for each of the relations in Exercise 6.6 and find which of these inverse relations are reflexive, symmetric, transitive, or antisymmetric.

6.8 Which of the relations in Exercises 6.6 and 6.7 are functions?

6.9 Determine the following compositions of relations defined in Exercise 6.6.

(a) $R_1 \circ R_1$ (h) $R_6 \circ R_6^{-1}$

(b) $R_1 \circ R_1^{-1}$ (i) $R_1 \circ R_7$

(c) $R_1 \circ R_2$ (j) $R_2 \circ R_7$

(d) $R_3 \circ R_2^{-1}$ (k) $R_3 \circ R_7$

(e) $R_3 \circ R_3$ (l) $R_8 \circ R_8$

(f) $R_4 \circ R_5$ (m) $R_8 \circ R_8^{-1}$

(g) $R_6 \circ R_6$ (n) $R_9 \circ R_{10}$

6.10 Which of the following statements are true provided that R, R_1, R_2 are binary relations defined on $A \times A$, where A is a finite set?

(a) If R is reflexive, then R^{-1} is reflexive

(b) If R is symmetric, then $R^{-1} = R$

(c) If R is transitive, then R^{-1} is transitive

(d) If both R_1 and R_2 are reflexive, then $R_1 \cap R_2$ is reflexive

(e) If both R_1 and R_2 are reflexive, then $R_1 \cup R_2$ is reflexive

(f) If both R_1 and R_2 are symmetric, then $R_1 \cap R_2$ is symmetric

(g) If both R_1 and R_2 are symmetric, then $R_1 \cup R_2$ is symmetric

(h) If both R_1 and R_2 are transitive, then $R_1 \cap R_2$ is transitive

(i) If both R_1 and R_2 are transitive, then $R_1 \cup R_2$ is transitive

(j) If both R_1 and R_2 are not symmetric, then $R_1 \cup R_2$ is not symmetric

(k) If R is an equivalence relation, then $R \circ R^{-1} = R$

6.11 Let $A_1, A_2, ..., A_n$ be finite sets.

(a) How many elements are contained in the Cartesian product $A_1 \times A_2 \times ... \times A_n$?

(b) How many different n-dimensional relations can be defined on this Cartesian product?

6.12 For each of the following functions, determine the supremum and infimum, as well the maximum and minimum (if they exist).

(a) $f(x) = 5$

(b) $f(x) = 5 - x, x \in [1, 5)$

(c) $f(x) = \sin(x), x \in [0, 2\pi]$

(d)
$$f(x) = \begin{cases} x & \text{when} & x \in [0, 1) \\ 1 - x & \text{when} & x \in [1, 2] \\ 0 & \text{otherwise} \end{cases}$$

(e) $f(x) = \dfrac{x}{1 + x}, x \in (0, 10)$

(f) $f(x) = \dfrac{\sin x}{x}, x \in \mathbb{R}$

6.13 For each of the following relations in the two-dimensional Euclidean space, determine its one-dimensional projections, its cylindric extensions, and the intersection of the cylindric extensions. In each case, show also a geometrical interpretation of the relation and the other relations derived from it.

(a) $R = \{\langle x, y \rangle | z^2 + y^2 \le 1\}$

(b) $R = \{\langle x, y \rangle | x \in [-1, 1], y \in [-1, 1]\}$

(c) $R = \{\langle x, y \rangle | |x| + |y| \le 1\}$

(d) $R = \{\langle x, y \rangle | z^2 + 2y^2 \le 1\}$

7

FUZZY RELATIONS

7.1 INTRODUCTION

While classical relations describe solely the *presence or absence* of association (interaction, connection, etc.) between elements of two or more sets, fuzzy relations are capable of capturing the strength of association (interaction, connection). In general, *fuzzy relations* are fuzzy sets defined on universal sets which are Cartesian products.

Consider, as an example, a binary fuzzy relation R defined on a set D of documents and a set T of key terms, which is important in information retrieval systems. Its membership function is defined on the Cartesian product $D \times T$. For each document d in set D and each key term t in set T, the membership degree $R(d, t)$ may be interpreted in this case as the *degree of relevance* of the document d to the key term t. This characterization of the relationship between documents and key terms is highly expressive. Any crisp relation defined on the same Cartesian product is far less expressive: It requires that each document be declared as either relevant or not relevant to each key term.

Consider the crisp equality relation E between variables x and y that is illustrated in Fig. 6.2. This relation captures the concept "x is equal to y." To capture a broader concept, expressed in linguistic terms as "x is approximately equal to y" or "x is close to y," we need a fuzzy relation. It is reasonable to capture this concept by the membership function

$$E(x, y) = \max\left(0, 1 - \frac{|x - y|}{c}\right)$$

145

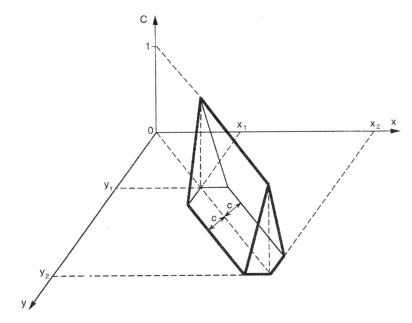

Figure 7.1   Membership function of the fuzzy relation "y is approximately equal to x."

where c is a positive real number whose value is chosen in the context of each application with the aim to make the fuzzy relation E a good representation of the concept in that application. This membership function is visually depicted in Fig. 7.1. As in Fig. 6.2, it is assumed here that the ranges of the variables are $[x_1, x_2]$ and $[y_1, y_2]$. The universal set on which this fuzzy relation is defined is thus the Cartesian product $[x_1, x_2] \times [y_1, y_2]$.

We can see that, in comparison with crisp relations, fuzzy relations substantially enhance our capability of dealing with relational concepts expressed in natural language. The purpose of this chapter is to explain how the various concepts pertaining to crisp relations introduced in Chapter 6 are extended to their fuzzy counterparts.

7.2 REPRESENTATIONS

A finite fuzzy relation can always be represented by a list of ordered pairs (or n-tuples) of the relevant Cartesian products with their membership grades. Pairs (or n-tuples) whose membership grades are zero are usually omitted.

When a fuzzy relation is not finite and involves n-dimensional Euclidean space for some $n \geq 2$, we have to define its membership function by a suitable formula. An example is the fuzzy relation "x is approximately equal to y," defined in the two-dimensional Euclidean space, whose membership function E is introduced in Sec. 7.1 and illustrated in Fig. 7.1.

Several other representations of finite fuzzy relations, which have some advantages over the list representation, are reviewed in this section.

Matrices

The most fundamental way of representing finite fuzzy relations is by matrices or, more generally, n-dimensional arrays, whose entries are membership grades corresponding to the individual n-tuples of the Cartesian products involved. To establish a unique correspondence between the n-tuples and the entries in the array, it is essential that some ordering of elements be defined for each set taking part in the Cartesian product. Consider a binary fuzzy relation R on $X \times Y$, where $X = \{x_1, x_2, \dots, x_n\}$ and $Y = \{y_1, y_2, \dots, y_m\}$. Then, using these orderings of the elements in sets X and Y, we define a unique matrix representation $\mathbf{R}$ of this relation as follows:

$$\mathbf{R} = \begin{bmatrix} r_{11} & r_{12} & \cdots & r_{1m} \\ r_{21} & r_{22} & \cdots & r_{2m} \\ \cdots & \cdots & \cdots & \cdots \\ r_{n1} & r_{n2} & \cdots & r_{nm} \end{bmatrix}$$

where $r_{ij} = R(x_i, y_j)$ for each $i = 1, 2, \dots, n$ and $j = 1, 2, \dots, m$. That is, the entry in the ith column and jth row of the matrix (r_{ij}) represents the membership degree $R(x_i, y_j)$ of pair (x_i, y_j) in fuzzy relation R.

If it is not desirable or practical to define a fixed ordering of elements of the sets involved in a given relation, the rows and columns of the matrix must be labeled by the elements they represent. As an example, let X denote a set of eight major cities,

$X = \{$Beijing, Chicago, London, Moscow, New York, Paris, Sydney, Tokyo$\}$

and let R denote a fuzzy relation on X that attempts to capture the relational concept *very far from*. A reasonable version of this relation is concisely represented by the matrix

R	B	C	L	M	N	P	S	T
B	0	1	0.7	0.5	1	0.7	0.6	0.1
C	1	0	0.5	0.9	0	0.5	1	1
L	0.7	0.5	0	0.3	0.5	0	1	0.7
M	0.5	0.9	0.3	0	0.9	0.3	0.8	0.5
N	1	0	0.5	0.9	0	0.5	1	1
P	0.7	0.5	0	0.3	0.5	0	1	0.7
S	0.6	1	1	0.8	1	1	0	0.6
T	1	1	0.7	0.5	1	0.7	0.6	0

in which the cities are abbreviated by their first letters.

To represent a three-dimensional relation, a sequence of matrices (referred to as a three-dimensional array) is needed. While two of the three dimensions are represented by columns and rows of the matrices, elements of the third dimension are represented by the distinct matrices. According to a common practice, we denote the underlying Cartesian product by $X_3 \times X_2 \times X_1$. Then, elements of X_1 correspond to columns in the matrices, elements of X_2 correspond to rows in the matrices, and elements of X_3 correspond to distinct matrices. Similarly, a sequence of three-dimensional arrays is needed to represent a four-dimensions relation, where the underlying Cartesian product is $X_4 \times X_3 \times X_2 \times X_1$. The distinct three-dimensional arrays now correspond with the elements of X_4. Using the same principle, arrays for higher dimensional relations are constructed.

Mappings

The visual representations of crisp binary relations on finite Cartesian products, called mappings, can be easily extended to fuzzy relations. We simply attach to each connection in the mapping diagram the membership grade of the respective ordered pair.

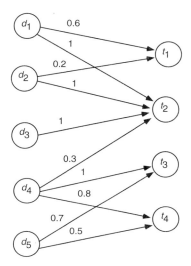

Figure 7.2   Representation of a fuzzy relation by the mapping diagram.

Consider as an example a set of documents $D = \{d_1, d_2, ..., d_5\}$, and a set of key terms $T = \{t_1, t_2, t_3, t_4\}$. Then a fuzzy relation expressing the degree of relevance of each document to each key term can be represented in the form of a diagram as shown in Fig. 7.2.

Directed Graphs

Directed graphs, introduced in Sec. 6.2 for finite crisp relations on X, can also be easily extended to fuzzy relations. To each directed connection between nodes in a graph, we simply attach the membership grade of the respective ordered pair.

The graphical representation of a fuzzy relation on X, where $X = \{x_1, x_2, x_3, x_4\}$, is shown in Fig. 7.3.

7.3   OPERATIONS ON BINARY FUZZY RELATIONS

Since fuzzy relations are special fuzzy sets (subsets of Cartesian products), all operations on fuzzy sets (complements, intersections,

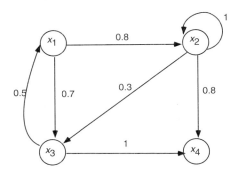

Figure 7.3   Representation of a fuzzy
relation by a directed graph.

unions, etc.) are applicable to fuzzy relations as well. We can also
apply the concept of subset and other concepts introduced for fuzzy
sets (sigma count, convexity, α-cuts, etc.) to fuzzy relations.

The focus of this section is on those operations on binary fuzzy
relations that are not applicable to ordinary fuzzy sets. These are the
operations of inverse and composition of fuzzy binary relations. They
are extensions of their crisp counterparts introduced in Sec. 6.3.

The *inverse* of a fuzzy binary relation R on $X \times Y$, which is usually denoted by R^{-1}, is a relation on $Y \times X$ defined as

$$R^{-1}(y, x) = R(x, y)$$

for all pairs $\langle y, x \rangle \in Y \times X$. Clearly,

$$(R^{-1})^{-1} = R$$

holds for any fuzzy binary relation.

When R is represented by a matrix, we obtain the matrix representation of R^{-1} by exchanging the rows of the given matrix with the
columns. The resulting matrix is called the *transpose* of the given
matrix. Thus, for example, if we denote the relation given in Fig. 7.2
by R, we have

$$
\mathbf{R} = \begin{bmatrix} 0.6 & 1 & 0 & 0 \\ 0.2 & 1 & 0 & 0 \\ 0 & 1 & 0 & 0 \\ 0 & 0.3 & 1 & 0.8 \\ 0 & 0 & 0.7 & 0.5 \end{bmatrix} \quad \text{and } \mathbf{R}^{-1} = \begin{bmatrix} 0.6 & 0.2 & 0 & 0 & 0 \\ 1 & 1 & 1 & 0.3 & 0 \\ 0 & 0 & 0 & 1 & 0.7 \\ 0 & 0 & 0 & 0.8 & 0.5 \end{bmatrix}
$$

As explained in Sec. 6.3, the composition of two crisp binary relations P and Q requires that the relations be compatible, which means that their Cartesian products $X \times Y$ and $Y \times Z$ must share the set Y. The composition $R = P \circ Q$ consists of those pairs $\langle x, z \rangle$ of the Cartesian product $X \times Z$ that can be connected via the two given relations P and Q and at least one element y in Y.

When P and Q are fuzzy relations, each connection from x to z via the relations and a particular element $y \in Y$ is a matter of degree. This degree depends on the membership degrees $P(x, y)$ and $Q(y, z)$ and is determined by the smaller of these two membership degrees. That is, the membership degree of a chain $\langle x, y, z \rangle$ is determined by the degree of the weaker of the two links, $\langle x, y \rangle$ and $\langle y, z \rangle$, in that chain. For example, if one of the two degrees is zero, then, regardless of the other degree, the degree of the chain should be zero. Also, among the chains that connect x to z, the largest degree of membership should be the membership degree which characterizes the relationship of x to z. Hence, the composition of fuzzy relations P and Q is defined for each pair $\langle x, z \rangle \in X \times Z$ by the formula

$$
R(x, z) = (P \circ Q)(x, z) = \max_{y \in Y} \ \min \ [P(x, y), Q(y, z)] \tag{7.1}
$$

The composition of fuzzy relations is easier to visualize in terms of fuzzy relational mappings. As an example, let

$X = \{a, b, c\}$
$Y = \{1, 2, 3, 4\}$
$Z = \{A, B, C\}$

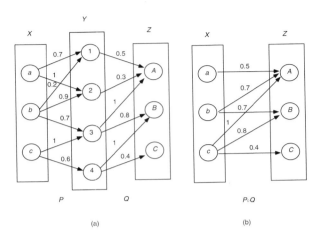

Figure 7.4   Illustration of the composition of binary fuzzy relations P and Q.

and consider the fuzzy relations P, Q defined on $X \times Y$ and $Y \times Z$ by their relational mappings in Fig. 7.4a. The composition of the relations is given in Fig. 7.4b. For example,

$$
\begin{aligned}
(P \circ Q)\,(b, A) &= \max\,\{\min[P(b, 1), Q(1, A)],\ \min[P(b, 2), Q(2, A)],\\
&\qquad \min\,[P(b, 3), Q(3, A)]\}\\
&= \max\,\{\min\,[0.2, 0.5],\ \min\,[0.9, 0.3],\ \min[0.7, 1]\}\\
&= \max\,\{0.2, 0.3, 0.7\} = 0.7
\end{aligned}
$$

Compositions of fuzzy relations are conveniently performed in terms of their matrix representations. Let

$$\mathbf{P} = [p_{ij}],\ \mathbf{Q} = [q_{jk}],\ \text{and}\ \mathbf{R} = [r_{ik}]$$

be matrix representations of fuzzy relations for which $P \circ Q = R$. Then, by using matrix notation, we can write

$$[r_{ik}] = [p_{ij}] \circ [q_{jk}]$$

where

$$r_{ik} = \max_{j}\ \min\,(p_{ij}, q_{jk})$$

Observe that the same entries in matrices **P** and **Q** are used to calculate matrix **R** as would be used in the regular matrix multiplication, but the product and sum are replaced here with the min and max operations, respectively.

To illustrate the use of matrices for computing compositions of fuzzy relations, let

$$X = \{p_1, p_2, p_3, p_4\}$$
$$Y = \{s_1, s_2, s_3\}$$
$$Z = \{d_1, d_2, d_3, d_4, d_5\}$$

be a set of patients, a set of symptoms, and a set of diseases, respectively. Assume that a fuzzy relation P on $X \times Y$ is defined by the matrix

$$\mathbf{P} = \begin{bmatrix} 0 & 0.3 & 0.4 \\ 0.2 & 0.5 & 0.3 \\ 0.8 & 0 & 0 \\ 0.7 & 0.7 & 0.9 \end{bmatrix}$$

This relation, which is obtained by examining the patients, describes how strongly the symptoms are manifested in the patients. Assume further that another fuzzy relation Q is defined on $Y \times Z$ by the matrix

$$\mathbf{Q} = \begin{bmatrix} 0.7 & 0 & 0 & 0.3 & 0.6 \\ 0.5 & 0.5 & 0.8 & 0.4 & 0 \\ 0 & 0.7 & 0.2 & 0.9 & 0 \end{bmatrix}$$

This relation describes a segment of medical knowledge expressing how strongly each symptom is associated with a disease. By performing the composition $P \circ Q = R$ in matrix form, we obtain

$$\begin{bmatrix} 0 & 0.3 & 0.4 \\ 0.2 & 0.5 & 0.3 \\ 0.8 & 0 & 0 \\ 0.7 & 0.7 & 0.9 \end{bmatrix} \circ \begin{bmatrix} 0.7 & 0 & 0 & 0.3 & 0.6 \\ 0.5 & 0.5 & 0.8 & 0.4 & 0 \\ 0 & 0.7 & 0.2 & 0.9 & 0 \end{bmatrix} = \begin{bmatrix} 0.3 & 0.4 & 0.3 & 0.4 & 0 \\ 0.5 & 0.5 & 0.5 & 0.4 & 0.2 \\ 0.7 & 0 & 0 & 0.3 & 0.6 \\ 0.7 & 0.7 & 0.7 & 0.9 & 0.6 \end{bmatrix}$$

For example,

$$0.3 \ (= r_{11}) = \max \ [\min(p_{11}, q_{11}), \min(p_{12}, q_{21}), \min(p_{13}, q_{31})]$$
$$= \max \ [\min(0,0.7), \min(0.3,0.5), \min(0.4, 0)]$$
$$= \max \ [0,0.3, 0]$$
$$0.7 \ (= r_{43}) = \max \ [\min(p_{41}, q_{13}), \min(p_{42}, q_{23}), \min(p_{43}, q_{33})]$$
$$= \max \ [\min(0.7, 0), \min(0.7, 0.8), \min(0.9,0.2)]$$
$$= \max \ [0, 0.7, 0.2]$$

The composite relation $\mathbf{R}$ in this example expresses the association between patients and diseases and, hence, facilitates medical diagnosis.

The two basic properties introduced in Sec. 6.3 for compositions of crisp relations are valid for compositions of fuzzy relations as well. That is,

$$(P \circ Q) \circ R = P \circ (Q \circ R)$$

and

$$(P \circ Q)^{-1} = Q^{-1} \circ P^{-1}$$

provided that the fuzzy relations P, Q, R are compatible. Also, in general,

$$P \circ Q \neq Q \circ P$$

for fuzzy relations.

7.4 FUZZY EQUIVALENCE RELATIONS AND COMPATIBILITY RELATIONS

The concept of an equivalence relation, introduced and examined in Sec. 6.4, can readily be fuzzified by properly reformulating the three properties that characterize equivalence relations: reflexivity, symmetry, and transitivity. The reformulations are trivial for reflexivity and symmetry: We say that a fuzzy relation R on X is *reflexive* if and only if $R(x,x) = 1$ for all $x \in X$; it is *symmetric* if and only if $R(x,y) = R(y,x)$ for all $x, y \in X$.

The property of *fuzzy transitivity* is not so obvious. Fuzzy transitivity can be defined in numerous ways, all of which collapse to the classical definition of transitivity for crisp relations. According to the most common definition, a fuzzy relation R is transitive if and only if

$$R(x, z) \geq \max_{y \in Y} \min[R(x, y), R(y, z)] \tag{7.2}$$

for all $x, z \in X$.

Observe that the formula on the right-hand side of this inequality expresses the composition $R \circ R$ of the relation R with itself. This is possible since R is defined on the Cartesian product $X \times X$ and, hence, it is compatible with itself. By performing the composition $R \circ R$, we obtain for each pair $\langle x, z \rangle \in X^2$ its membership grade representing the indirect connection of elements x and z via all possible chains with two links. For a fuzzy relation to be transitive (i.e., to satisfy the inequality), it is required that for any pair $\langle x, z \rangle \in R$, the direct membership grade $R(x, z)$ be not smaller than the membership grade obtained indirectly.

A fuzzy relation on X that is reflexive, symmetric, and transitive in the sense of the introduced definitions is a *fuzzy equivalence relation*. It is easy to verify that this conception of fuzzy equivalence is cutworthy. That is, each α-cut of any fuzzy equivalence relation is an equivalence relation in the classical sense. Moreover, as α is increased, equivalence classes in the α-cuts become more refined.

As a simple example, consider a set of six experts identified by numbers 1 through 6, who are asked to express their opinions on some policy issue. Assume that the fuzzy relation Q expressed on this set by the matrix

$$\mathbf{Q} = \begin{bmatrix} 1 & 0.8 & 0 & 0.8 & 0.5 & 0 \\ 0.8 & 1 & 0 & 1 & 0.5 & 0 \\ 0 & 0 & 1 & 0 & 0 & 0.8 \\ 0.8 & 1 & 0 & 1 & 0.5 & 0 \\ 0.5 & 0.5 & 0 & 0.5 & 1 & 0 \\ 0 & 0 & 0.8 & 0 & 0 & 1 \end{bmatrix}$$

attempts to capture the degree of similarity of opinions on the issue in question for each pair of experts. This relation is reflexive, symmetric, and transitive. As such, it is a fuzzy equivalence relation. A diagram of this relation is shown in Fig. 7.5a. This is a simplified diagram in

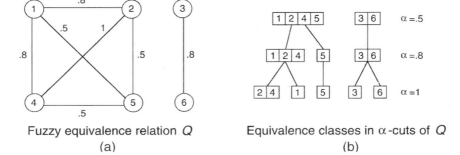

Fuzzy equivalence relation Q Equivalence classes in α-cuts of Q
          (a)   (b)

Figure 7.5   An example of fuzzy equivalence relation.

which the connection from each node to itself (due to reflexivity) is
omitted and connections between nodes are not directed (due to symmetry). We can see that each α-cut of this fuzzy relation is a crisp
equivalence relation. Partitions induced by the equivalence classes of
α-cuts for $\alpha = 0.5, 0.8, 1$ are shown in Fig. 7.5b. It is significant that
these partitions become more refined when the value of α is increased.

Analogous to crisp compatibility relations, *fuzzy compatibility*
relations do not require transitivity. They are only reflexive and symmetric in the same sense as fuzzy equivalence relations. Each fuzzy
compatibility relation is cutworthy.

As an example, consider again the set of experts, labeled 1
through 6, who express their opinions on a given policy issue. Assume
that the similarity in their opinions is captured by the fuzzy relation R
expressed by the matrix

$$\mathbf{R} = \begin{bmatrix} 1 & 1 & 0 & 0.8 & 0.9 & 0 \\ 1 & 1 & 0.8 & 0.9 & 0.5 & 0 \\ 0 & 0.8 & 1 & 0 & 0 & 0.8 \\ 0.8 & 0.9 & 0 & 1 & 1 & 0 \\ 0.9 & 0.5 & 0 & 1 & 1 & 0 \\ 0 & 0 & 0.8 & 0 & 0 & 1 \end{bmatrix}$$

or by the simplified diagram in Fig. 7.6a. It is easy to see that this
relation is reflexive and symmetric, but not transitive. For example,

$$R(1, 4) < \max \{\min [R(1, 2), R(2, 4)], \min [R(1, 5), R(5, 4)]\}$$

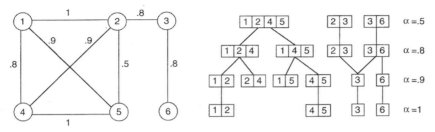

Fuzzy compatibility relation R Maximal compatibility classes in α-cuts of R
    (a)   (b)

Figure 7.6   An example of fuzzy compatibility.

which violates the property of transitivity. Maximal compatibility classes induced by the crisp compatibility relations at distinct α-cuts of R are shown in Fig. 7.6 b. We can see that the compatibility classes become more refined when the value of α is increased.

The last two examples illustrate that similarity among elements of some set with respect to some attribute may be expressed either by a fuzzy equivalence relation or by a fuzzy compatibility relation. The former conception of similarity is qualitatively stronger than the latter.

In some applications, a fuzzy relation that should be transitive on intuitive grounds is actually not transitive. This unsatisfactory situation may be caused by a deficiency in the data from which the relation was derived, inconsistent opinions of experts, or other shortcomings. Transitivity is essential, for example, if the relation is intuitively an equivalence relation (as in some problems of fuzzy clustering) or an ordering relation (as in various problems of decision making). In such cases, it is desirable to convert the given fuzzy relation R to a transitive one that is as close as possible to R. Such a relation is called the *transitive closure* of R. To obtain transitivity, some degrees of membership in R must be properly increased. The transitive closure of R is thus the smallest fuzzy relation that is transitive and contains R.

The transitive closure, R_T, of a fuzzy relation R can be determined by a simple iterative algorithm that consists of the following two steps:

(1) Compute $R' = R \cup (R \circ R)$;
(2) If $R' \neq R$, rename R' as R and go to step (1); otherwise $R' = R_T$ and the algorithm terminates.

Assuming that the given relation R is defined on a set with n elements, the algorithm guarantees that we obtain the transitive closure in less than $n - 1$ iterations.

To illustrate the use of this algorithm, let us calculate the transitive closure R_T of the compatibility relation R in Fig. 7.6. The following are the three matrices participating in step (1) of the first iteration of the algorithm:

$$
\begin{bmatrix}
1 & 1 & 0 & 0.8 & 0.9 & 0 \\
1 & 1 & 0.8 & 0.9 & 0.5 & 0 \\
0 & 0.8 & 1 & 0 & 0 & 0.8 \\
0.8 & 0.9 & 0 & 1 & 1 & 0 \\
0.9 & 0.5 & 0 & 1 & 1 & 0 \\
0 & 0 & 0.8 & 0 & 0 & 1
\end{bmatrix}
\begin{bmatrix}
1 & 1 & 0.8 & 0.9 & 0.9 & 0 \\
1 & 1 & 0.8 & 0.9 & 0.9 & 0.8 \\
0.8 & 0.8 & 1 & 0.8 & 0.5 & 0.8 \\
0.9 & 0.9 & 0.8 & 1 & 1 & 0 \\
0.9 & 0.9 & 0.5 & 1 & 1 & 0 \\
0 & 0.8 & 0.8 & 0 & 0 & 1
\end{bmatrix}
\begin{bmatrix}
1 & 1 & 0.8 & 0.9 & 0.9 & 0 \\
1 & 1 & 0.8 & 0.9 & 0.9 & 0.8 \\
0.8 & 0.8 & 1 & 0.8 & 0.5 & 0.8 \\
0.9 & 0.9 & 0.8 & 1 & 1 & 0 \\
0.9 & 0.9 & 0.5 & 1 & 1 & 0 \\
0 & 0.8 & 0.8 & 0 & 0 & 1
\end{bmatrix}
$$

$$
\qquad\qquad R \qquad\qquad\qquad\qquad R \circ R \qquad\qquad\qquad R \cup (R \circ R)
$$

We can see that $R \circ R = R \cup (R \circ R)$ in this case. In general, however,

$$
R \circ R \subseteq R \cup (R \circ R)
$$

Next, we have to perform step (2) of the algorithm: Since $R' \neq R$, we rename R' as R and repeat step (1). After the second iteration the resulting matrix is:

$$
\begin{bmatrix}
1 & 1 & 0.8 & 0.9 & 0.9 & 0.8 \\
1 & 1 & 0.8 & 0.9 & 0.9 & 0.8 \\
0.8 & 0.8 & 1 & 0.8 & 0.8 & 0.8 \\
0.9 & 0.9 & 0.8 & 1 & 1 & 0.8 \\
0.9 & 0.9 & 0.8 & 1 & 1 & 0.8 \\
0.8 & 0.8 & 0.8 & 0.8 & 0.8 & 1
\end{bmatrix}
$$

$$
R \cup (R \circ R)
$$

Since this relation (denoted by R') is different from the relation obtained in the first iteration (denoted now as R), we again rename R' as R and repeat step (1). As the reader can check, the relation now does not change. This means that we have obtained the sought transitive closure and the algorithm terminates. The diagram of the transi-

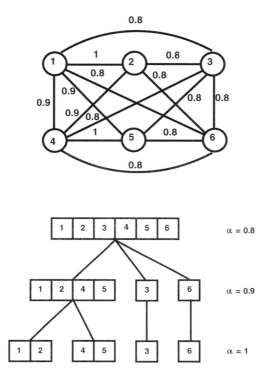

Figure 7.7   Equivalence relation that is the transitive closure of the compatibility relation in Fig. 7.6.

tive closure R_T and the equivalence classes of the three distinct α-cuts of R_T are shown in Fig. 7.7.

Observe that the described algorithm can be used not only for calculating transitive closures of given fuzzy relations, but also for verifying transitivity of fuzzy relations that are supposed to be transitive. If a given relation is transitive and we apply the algorithm to it, the algorithm terminates after the first iteration.

7.5 FUZZY PARTIAL ORDERINGS

Analogous to crisp partial orderings, fuzzy partial orderings are fuzzy relations on X that satisfy the fuzzy counterparts of reflexivity, transitivity, and antisymmetry. Fuzzy reflexivity and transitivity were already introduced in the context of fuzzy equivalence relations. The reformulation of antisymmetry for fuzzy relations is trivial: A fuzzy relation R on X is *antisymmetric* when $R(x, y) > 0$ and $R(y, x) > 0$ imply that $x = y$ for any $x, y \in X$. It is clear that this conception of fuzzy antisymmetry is cutworthy. Since fuzzy reflexivity and fuzzy transitivity, as defined in Sec. 7.4, are also cutworthy, fuzzy partial orderings are cutworthy as well.

Fuzzy partial orderings have broad utility. They can be applied, for example, when expressing our preferences within a set of alternatives. Compared with crisp partial orderings, they have greater expressive power. They allow us to express not only that we prefer one alternative to another, but also the strength of this preference.

As a simple example, consider the set $X = \{x_1, x_2, x_3, x_4, x_5\}$ of alternatives. Assume that our preference regarding these five alternatives are expressed by the matrix

$$\mathbf{R} = \begin{bmatrix} 1 & 0 & 0.5 & 0 & 0 \\ 0.7 & 1 & 0.9 & 0 & 0.1 \\ 0 & 0 & 1 & 0 & 0 \\ 1 & 0.9 & 1 & 1 & 0.9 \\ 0.7 & 0 & 0.8 & 0 & 1 \end{bmatrix}$$

According to this matrix, for example, x_1 is preferred to x_3 with the degree of 0.5, x_3 is preferred to x_1 with the degree 0, x_4 is preferred to x_5 with the degree of 0.9, and so forth. In general, we say that x_i is preferred to x_j if $x_i \leq x_j$ according to R. By convention, we define the degree of preference of each alternative to itself as 1. Examining the matrix, we can see that it represents a fuzzy partial ordering: It is reflexive (by our convention), antisymmetric, and transitive (as can be easily verified by the algorithm for transitive closure introduced in Sec. 7.4). A diagram of this relation and the Hasse diagrams of its α-cuts are shown in Fig. 7.8, where the five alternatives x_1, x_2, x_3, x_4, x_5 are denoted by their subscripts. As expected, all α-cuts are crisp partial orderings. With increasing values of α, these partial orderings become weaker. For $\alpha = 0.1$, the relation is linear, which means that all alternatives are comparable. For $\alpha = 0.5$, alternatives 2 and 5 are not comparable (i.e., we are indifferent to them at that level according to our expression of preferences). The number of pairs that are not comparable increases with increasing values of α. For $\alpha = 1$, no pairs are comparable except two: 4 is preferable to both 1 and 3.

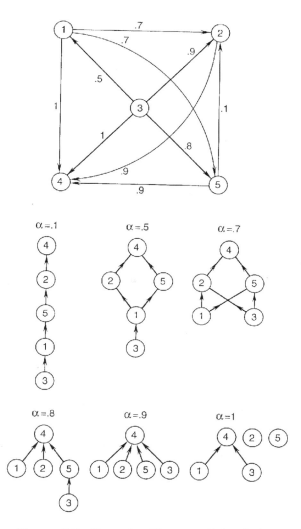

Figure 7.8   Example of fuzzy partial ordering.

When degrees of preferences are properly expressed, they should result in a fuzzy partial ordering. That is, it should be a fuzzy relation that is reflexive, antisymmetric, and transitive. The first two properties usually do not involve any difficulties and can easily be verified. However, due to a human error or some other reason, a fuzzy relation that is supposed to express some preference ordering may not be transitive. In such a case, we can convert the relation to its transitive closure by the algorithm introduced in Sec. 7.4.

7.6 PROJECTIONS AND CYLINDRIC EXTENSIONS

To facilitate our discussion of projections and cylindric extensions of fuzzy relations, let us consider the relation

$$\mathbf{Q} = \begin{bmatrix} 0.7 & 0 & 0 & 0.3 & 0.6 \\ 0.5 & 0.5 & 0.8 & 0.4 & 0 \\ 0 & 0.7 & 0.2 & 0.9 & 0 \end{bmatrix}$$

introduced in Sec. 7.3, which expresses the association of a set of symptoms, $S = \{s_1, s_2, s_3\}$ with a set of diseases, $D = \{d_1, d_2, d_3, d_4, d_5\}$.

Observe that each row and each column in the matrix representing the relation can be viewed as a representation of an ordinary fuzzy set. For example, the first row in $\mathbf{Q}$ represents the fuzzy set of diseases associated with symptom s_1, while the first column in $\mathbf{Q}$ represents the set of symptoms associated with disease d_1. The whole relation can thus be viewed as a collection of two families of ordinary fuzzy sets. If desirable, some fuzzy sets in either family may be aggregated by an appropriate aggregation operation. When all of them are aggregated by the operation of fuzzy union, we obtain a projection of the relation on the respective dimension. That is, the projection of our relation Q on the first dimension (set S), denoted by Q_1, is defined for each $s \in S$ by the formula

$$Q_1(s) = \max_{d \in D} Q(s, d)$$

Similarly, the projection of Q on the second dimension (set D), denoted by Q_2, is defined for each $d \in D$ by the formula

$$Q_2(d) = \max_{s \in S} Q(s, d)$$

We obtain,

$$Q_1 = 0.7/s_1 + 0.8/s_2 + 0.9/s_3$$
$$Q_2 = 0.7/d_1 + 0.7/d_2 + 0.8/d_3 + 0.9/d_4 + 0.6/d_6$$

Given an arbitrary n-dimensional fuzzy relation R on $X = X_1 \times X_2 \times ... \times X_n$, its projection can be defined with respect to any chosen subset of the n dimensions. Let the Cartesian product of the chosen dimensions (i.e., some of the sets $X_1, X_2, ..., X_n$) be denoted by P and the Cartesian product of the remaining dimensions by $\bar{P}$, and let each n-tuple $x \in X$ be written in terms of its components $p \in P$ and $\bar{p} \in \bar{P}$. Then, the projection R_P of R with respect to the dimensions employed in P is expressed, for each $p \in P$, by the formula

$$R_P(p) = \max_{\bar{p} \in \bar{P}} R(p, \bar{p})$$

Given an n-dimensional fuzzy relation R on $X = X_1 \times X_2 \times ... \times X_n$, any $(n+k)$-dimensional relation $(k \geq 1)$ whose projection into the n dimensions of R yields R is called an *extension* of R. Among all extensions into some specified additional dimensions, the largest one is called a *cylindric extension* of R into the specified dimensions.

Let an extension of R be made with respect to the Cartesian product $Y = Y_1 \times Y_2 \times ... \times Y_h$. Then, the cylindric extension, ^{EY}R, of R into Y is defined by the equation

$$^{EY}R(x, y) = R(x)$$

for all $x \in X$ and all $y \in Y$. For example, cylindric extensions of the two projections Q_1 and Q_2 of our relation Q with respect to D and S, respectively, are

$$^{ED}\mathbf{Q_1} = \begin{bmatrix} 0.7 & 0.7 & 0.7 & 0.7 & 0.7 \\ 0.8 & 0.8 & 0.8 & 0.8 & 0.8 \\ 0.9 & 0.9 & 0.9 & 0.9 & 0.9 \end{bmatrix}, \quad {}^{ES}\mathbf{Q_2} = \begin{bmatrix} 0.7 & 0.7 & 0.8 & 0.9 & 0.6 \\ 0.7 & 0.7 & 0.8 & 0.9 & 0.6 \\ 0.7 & 0.7 & 0.8 & 0.9 & 0.6 \end{bmatrix}$$

The intersection of these cylindric extensions contains Q, but it is not equal to it:

$$^{ED}\mathbf{Q_1} \cap {}^{ES}\mathbf{Q_2} = \begin{bmatrix} 0.7 & 0.7 & 0.7 & 0.7 & 0.6 \\ 0.7 & 0.7 & 0.8 & 0.8 & 0.6 \\ 0.7 & 0.7 & 0.8 & 0.9 & 0.6 \end{bmatrix} \supseteq Q$$

This means that Q is not reconstructable from its two projections. In other words, it is not a cylindric closure of the projections.

An example of a binary fuzzy relation that is reconstructable from its projections is a relation defined on $X \times Y$ by the matrix

$$\mathbf{R} = \begin{bmatrix} 0.3 & 0.4 & 0.4 & 0.4 \\ 0.3 & 0.5 & 0.7 & 0.7 \\ 0.2 & 0.2 & 0.2 & 0.2 \\ 0.3 & 0.5 & 0.8 & 0.9 \end{bmatrix}$$

where $X = \{x_1, x_2, x_3, x_4\}$ and $Y = \{y_1, y_2, y_3, y_4\}$. Its projections are

$$R_1 = 0.4/x_1 + 0.7/x_2 + 0.2/x_3 + 0.9/x_4$$
$$R_2 = 0.3/y_1 + 0.5/y_2 + 0.8/y_3 + 0.9/y_4$$

The cylindric extensions are defined by the matrices

$$^{EY}\mathbf{R_1} = \begin{bmatrix} 0.4 & 0.4 & 0.4 & 0.4 \\ 0.7 & 0.7 & 0.7 & 0.7 \\ 0.2 & 0.2 & 0.2 & 0.2 \\ 0.9 & 0.9 & 0.9 & 0.9 \end{bmatrix} \quad ^{EX}\mathbf{R_2} = \begin{bmatrix} 0.3 & 0.5 & 0.8 & 0.9 \\ 0.3 & 0.5 & 0.8 & 0.9 \\ 0.3 & 0.5 & 0.8 & 0.9 \\ 0.3 & 0.5 & 0.8 & 0.9 \end{bmatrix}$$

Their intersection is the original relation R, and, hence, R is a cylindric closure of its projections.

EXERCISES

7.1 Find some examples of fuzzy relations encountered in daily life. For each binary fuzzy relation defined on a single universal set, find if it is reflexive, symmetric, antisymmetric, or transitive.

7.2 Let

$$\mathbf{R} = \begin{bmatrix} 0.6 & 0.3 & 0.1 \\ 0.8 & 0.9 & 0.2 \end{bmatrix} \quad \text{and} \quad \mathbf{S} = \begin{bmatrix} 0.5 & 0.8 \\ 0.8 & 0.9 \\ 0.5 & 0.1 \end{bmatrix}$$

be matrix representations of binary fuzzy relations. Calculate the following max-min compositions:

(a) $R \circ S$

(b) $\bar{R} \circ \bar{S}$

(c) $S^{-1} \circ R^{-1}$

   $(R \circ S) \circ S^{-1}$

(e) $R^{-1} \circ (R \circ S)$

(f) $((R \circ S) \circ S) \circ R$

7.3 Let R and S be the binary fuzzy relations defined in Exercise 7.2 and let

$$\mathbf{T} = \begin{bmatrix} 0.2 & 0.5 & 0.7 & 0.9 \\ 0.3 & 0.8 & 0.6 & 0.1 \end{bmatrix}$$

be a matrix representation of another fuzzy relation.

(a) Show that $(R \circ S) \circ T = R \circ (S \circ T)$.

(b) Prove that this equality (associativity of max-min composition) holds for any compatible relations R, S, T.

7.4 Let R and S be the binary fuzzy relations defined in Exercise 7.2 and let

$$\mathbf{Q} = \begin{bmatrix} 0.3 & 0.6 & 0.8 \\ 0.4 & 0.5 & 0.7 \end{bmatrix}$$

be a matrix representation of another binary fuzzy relation. Calculate the following binary fuzzy relations:

(a) $(R \cap S) \circ S$ and $(R \cup S) \circ S$

(b) $(R \cap Q \cap S^{-1}) \circ S$ and $(R \cup Q \cup S^{-1}) \circ S$

(c) $(R \circ S) \cap (Q \circ S)$ and $(R \circ S) \cup (Q \circ S)$

(d) $(R \cap Q) \circ S$ and $(R \cup Q) \circ S$

7.5 Let

$$\mathbf{P} = \begin{bmatrix} 1 & 0.6 & 0.9 \\ 0.4 & 0.8 & 0.7 \\ 1 & 0.9 & 1 \end{bmatrix}$$

be a matrix representation of a binary fuzzy relation. Calculate $P \circ P$, $P \circ P \circ P$, and $P \circ P \circ P \circ P$.

7.6 Represent the relations defined in Exercises 7.2–7.4 by mappings.

7.7 Represent the relation P (defined in Exercise 7.5) and the three relations obtained by the specified compositions in the exercise by directed graphs.

7.8 Given a binary fuzzy relation R on a finite set X, show that the relation is transitive if and only if $R = (R \circ R) \cup R$. (This property can be utilized as a convenient test of transitivity of binary fuzzy relations on X.)

7.9 Let

$$
R = \begin{bmatrix}
1 & 0.8 & 0 & 0.4 & 0 & 0 & 0 \\
0.8 & 1 & 0 & 0.4 & 0 & 0 & 0 \\
0 & 0 & 1 & 0 & 1 & 0.9 & 0.5 \\
0.4 & 0.4 & 0 & 1 & 0 & 0 & 0 \\
0 & 0 & 1 & 0 & 1 & 0.9 & 0.5 \\
0 & 0 & 0.9 & 0 & 0.9 & 1 & 0.5 \\
0 & 0 & 0.5 & 0 & 0.5 & 0.5 & 1
\end{bmatrix}
$$

be a matrix a representation of a binary fuzzy relation on $X = \{1, 2, 3, 4, 5, 6, 7\}$.
  (a) Show that the relation is reflexive, symmetric, and transitive.
  (b) Determine equivalence classes induced by this fuzzy equivalence relation in the α-cuts of R and express them in the form shown in Fig. 7.5b.

7.10 Modify the matrix in Exercise 7.9 by replacing the entries $r_{36} = r_{63} = 0.9$ with $r_{36} = r_{63} = 0.7$.
  (a) Show that the new relation is reflexive and symmetric, but not transitive.
  (b) Determine maximum compatibility classes induced by the fuzzy compatibility relation in the α-cuts of R and express them in the form shown in Fig. 7.6b.

7.11 Determine the transitive closure of the relation defined in Exercise 7.5.

7.12 Let

$$
R = \begin{bmatrix}
1 & 0.7 & 0 & 0.8 & 0 \\
0 & 1 & 0 & 0.7 & 1 \\
0.5 & 0.8 & 1 & 0 & 0 \\
0 & 0 & 0.5 & 1 & 0 \\
0 & 0 & 0.9 & 0.9 & 1
\end{bmatrix}
$$

be a matrix representation of binary fuzzy relation on $X = \{1, 2, 3, 4, 5\}$.
  (a) Show that the relation is reflexive and antisymmetric.
  (b) Determine whether the relation is transitive (see Exercise 7.8); if it is not transitive, calculate its transitive closure.
  (c) Determine the Hasse diagrams of all α-cuts of transitive closure of R.

7.13 For each of the relations defined in Exercises 7.2 and 7.3, determine the following:
  (a) Both projections.
  (b) Cylindric extensions of the projections.
  (c) Intersection of the cylindric extensions.

7.14 Repeat Exercise 7.13 for the fuzzy binary relation in the two-dimensional Euclidean space whose membership function is shown in Fig. 7.1.

7.15 Consider two independent variables x and y whose values are real numbers. Assume that the actual values of these variables are approximated by the trapezoidal-shape fuzzy numbers A and B shown in Fig. 7.9. Draw a picture of the membership function representing the intersection of the cylindric extensions of A and B.

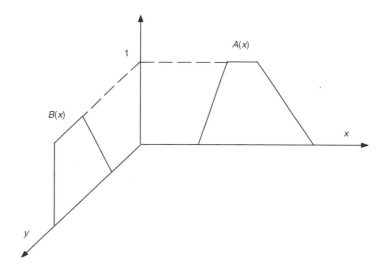

Figure 7.9 Illustration for Exercise 7.15.

8

FUZZY ARITHMETIC

8.1 FUZZY NUMBERS

The concept of a *fuzzy number* arises from the fact that many quantifiable phenomena do not lend themselves to being characterized in terms of absolutely precise numbers. For example, most of us have watches that are at least somewhat inaccurate, so we might say that the time is now "about two o' clock." Or, we may not wish to pin ourselves down to an exact schedule and, thus, issue an invitation to dinner for "around six-thirty." In a grocery store, we are satisfied if a bunch of bananas weighs "approximately four pounds." Thus, a fuzzy number is one which is described in terms of a number word and a linguistic modifier, such as *approximately, nearly,* or *around*.

Intuitively, we can see that the concept captured by the linguistic expression *approximately six* is fuzzy, because it includes some number values on either side of its central value of six. Although the central value is fully compatible with this concept, the numbers around the central value are compatible with it to lesser degrees. Intuitively, we feel that the degree of compatibility of each number with the concept should express, in some way dependent on the context, its proximity to the central value. That is, the concept can be captured by a fuzzy set defined on the set of real numbers. Its membership function should assign the degree of 1 to the central value and degrees to other numbers that reflect their proximity to the central value according to some rule. The membership function should thus decrease from

1 to 0 on both sides of the central value. Fuzzy sets of this kind are called *fuzzy numbers.*

It is not difficult to see that fuzzy numbers play an important role in many applications, including decision making, approximate reasoning, fuzzy control, and statistics with imprecise probabilities. We can imagine, for example, a decision-making situation in which a stock analyst concludes that if a particular stock reaches *about $50,* then the fund manager should sell *approximately half* of her available shares. Before we explore the implications of this concept, we must define the concept of a fuzzy number more precisely.

While every fuzzy number A is expressed by a membership function of the form

$$A: \mathbb{R} \to [0, 1]$$

not all membership functions of this form represent fuzzy numbers. To qualify as a fuzzy number, the membership function must capture our intuitive conception of a set of numbers that are around a given real number or, possibly, around an interval of real numbers. Membership functions that conform to this intuitive conception must be expressed in the general form

$$A(x) = \begin{cases} f(x) & \text{for } x \in [a,b] \\ 1 & \text{for } x \in [b,c] \\ g(x) & \text{for } x \in [c,d] \\ 0 & \text{for } x < a \text{ and } x > d \end{cases} \tag{8.1}$$

where $a \leq b \leq c \leq d$, f is a continuous function that increases to 1 at point b, and g is a continuous function that decreases from 1 at point c. Four membership functions that comply with this form are shown in Fig. 8.1.

While fuzzy sets C and D in Fig. 8.1 (characterized by $b = c$) conform to our intuitive conception of a fuzzy number, it seems more appropriate to view fuzzy sets A and B as fuzzy intervals. However, distinction is not usually made in the literature between fuzzy numbers and fuzzy intervals; both are subsumed under the common name "fuzzy number."

Although membership functions of a great variety of shapes are possible for representing fuzzy numbers, as exemplified in Fig. 8.2 for

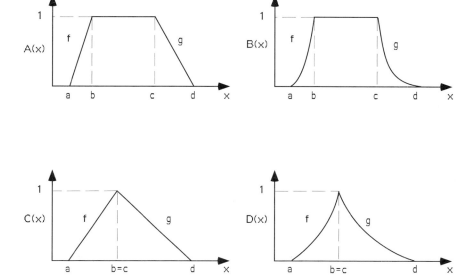

Figure 8.1   Examples of fuzzy numbers.

the concept "around 3," the most common are trapezoidal and triangular shapes. These types of fuzzy numbers are easy to construct and manipulate. Even though the choice of the real numbers $a,\ b,\ c,\ d$ in the general definition (8.1) of a fuzzy number is very important and highly dependent on the context of each application, most current applications that employ fuzzy numbers are not significantly affected by the shapes of functions f and g in (8.1). Hence, it is quite natural to choose simple linear functions, represented by straight lines.

When some of the real numbers $a,\ b,\ c,\ d$ in (8.1) are equal, we obtain various degenerate forms. They are illustrated for a trapezoidal-shaped fuzzy number in Fig. 8.3. All of these special fuzzy sets are viewed as fuzzy numbers. Observe also that we obtain a real number, as a very special case of a fuzzy number, when $a = b = c = d$.

We can easily see that the general form (8.1) required for fuzzy numbers implies the following properties of fuzzy numbers:

(a) Fuzzy numbers are normal fuzzy sets (i.e., the core of every fuzzy number is not empty).

(b) The α-cuts of every fuzzy number are closed intervals of real numbers.

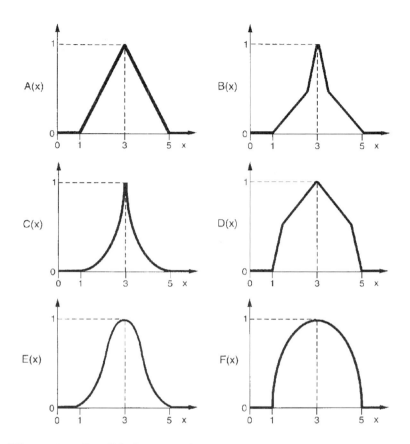

Figure 8.2   Possible fuzzy numbers to capture the concept *"around 3."*

(c) The support of every real number is the open interval (a, d) of real numbers.

(d) Fuzzy numbers are convex fuzzy sets.

These properties are essential for defining meaningful arithmetic operations on fuzzy numbers. Since each fuzzy set is uniquely represented by its α-cuts and these are closed intervals of real numbers, arithmetic operations on fuzzy numbers can be defined in terms of

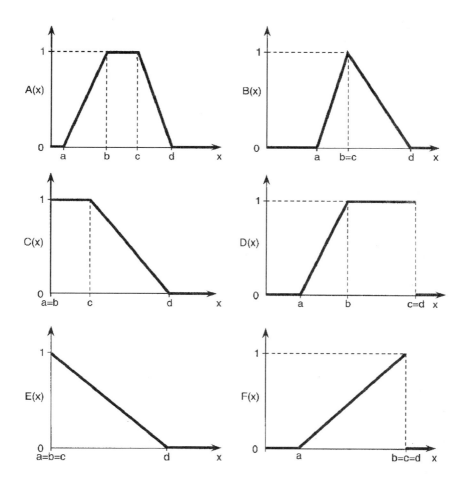

Figure 8.3   Trapezoidal-shaped fuzzy number and its various degenerated cases.

arithmetic operation on closed intervals of real numbers. These operations are the cornerstone of *interval analysis,* a well-established area of classical mathematics. We introduce them in the next section and then utilize them in Sec. 8.3 for defining arithmetic operations on fuzzy numbers.

8.2 ARITHMETIC OPERATIONS ON INTERVALS

In many application contexts, intervals of real numbers allow us to describe our uncertainty about the actual value of a numerical variable. This uncertainty may be caused, for example, by limited resolution of measuring instruments or by limited computing precision. Intervals of real numbers may also be used in a prescriptive way. For example, an income tax table may state that we owe $4,428 if our taxable income as a single person falls within the interval [$26,051, $26,100]. In decision making, intervals are often used for specifying acceptable or unacceptable values of relevant numerical variables. Similarly, in engineering design they are often used for characterizing acceptable variations of design parameters.

Mathematics for dealing with imprecise representations of real numbers in terms of closed intervals has been under development since the late 1950s, when its significance was first recognized. For our purpose in this book, we are only interested in the four basic arithmetic operations on closed intervals: addition, subtraction, multiplication, and division. To define these operations, let us consider two closed intervals $[a, b]$ and $[c, d]$. The endpoints of these intervals are some real numbers, denoted generically here as a, b, c, d, for which $a \leq b$ and $c \leq d$.

Given two closed intervals of real numbers, $[a, b]$ and $[c, d]$, the result of any of the four arithmetic operations on these intervals is defined as the set of real numbers obtained by performing the operation on each ordered pair of real numbers in the Cartesian product $[a, b] \times [c, d]$, except that the division $[a, b] / [c, d]$ is not defined when $0 \in [c, d]$. This is justified by the fact that each real number contained in either interval may be the actual number and hence, we have to perform the operation on all possible pairs of real numbers, one taken from $[a, b]$ and one from $[c, d]$, and take the set of these individual results as the overall result of the operation on the intervals.

We can easily see that the result of any arithmetic operation on closed intervals is also a closed interval. Moreover, the endpoints of the resulting interval can be expressed in terms of the endpoints of the given intervals $[a, b]$ and $[c, d]$.

We now introduce the four basic arithmetic operations on intervals in terms of their endpoints.

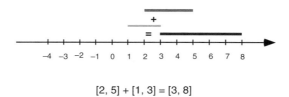

$$[2, 5] + [1, 3] = [3, 8]$$

$$[0, 1] + [-6, 5] = [-6, 6]$$

Figure 8.4   Interval addition.

Addition (+)

$$[a, b] + [c, d] = [a + c, b + d]$$

Examples (Fig. 8.4):

$$[2, 5] + [1, 3] = [2+1, 5+3] = [3, 8]$$

$$[0, 1] + [-6, 5] = [0-6, 1+5] = [-6, 6]$$

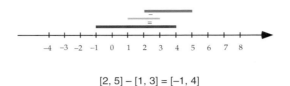

$$[2, 5] - [1, 3] = [-1, 4]$$

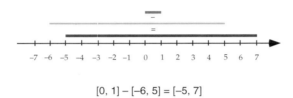

$$[0, 1] - [-6, 5] = [-5, 7]$$

Figure 8.5   Interval subtraction.

Subtraction (–)

$$[a, b] - [c, d] = [a{-}d, b{-}c]$$

Examples (Fig. 8.5):

$$[2, 5] - [1, 3] = [2{-}3, 5{-}1] = [-1, 4]$$

$$[0, 1] - [-6, 5] = [0{-}5, 1{+}6] = [-5, 7]$$

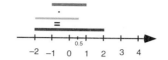

$$[-1, 1] \cdot [-2, 0.5] = [-2, 2]$$

$$[3, 4] \cdot [2, 2] = [6, 8]$$

Figure 8.6   Interval multiplication.

Multiplication (·)

$$[a, b] \cdot [c, d] = [\min(ac, ad, bc, bd), \max(ac, ad, bc, bd)]$$

Examples (Fig. 8.6):

$$
\begin{aligned}
[-1, 1] \cdot [-2, 0.5] &= [\min(-1 \cdot (-2), -1 \cdot 0.5, 1 \cdot (-2), 1 \cdot 0.5), \\
&= \max(-1 \cdot (-2), -1 \cdot 0.5, 1 \cdot (-2), 1 \cdot 0.5)] \\
&= [\min(2, -0.5, -2, 0.5), \max(2, -0.5, -2, 0.5)] \\
&= [-2, 2]
\end{aligned}
$$

$$
\begin{aligned}
[3, 4] \cdot [2, 2] &= [\min(3 \cdot 2, 3 \cdot 2, 4 \cdot 2, 4 \cdot 2), \max(3 \cdot 2, 3 \cdot 2, 4 \cdot 2, 4 \cdot 2)] \\
&= [\min(6, 6, 8, 8), \max(6, 6, 8, 8)] \\
&= [6, 8]
\end{aligned}
$$

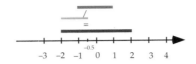

$$[-1, 1] / [-2, -0.5] = [-2, 2]$$

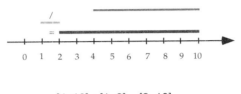

$$[4, 10] / [1, 2] = [2, 10]$$

Figure 8.7   Interval division.

Division (/)

$[a, b] / [c, d] = [a, b] \cdot [1/d, 1/c]$
$\quad\quad = [\min(a/c, a/d, b/c, b/d), \max(a/c, a/d, b/c, b/d)]$

   Interval division assumes that the number 0 is not one of the elements in the divisor interval $[c, d]$.

Examples (Fig. 8.7):

$[-1, 1] / [-2, -0.5] = [-1, 1] \cdot [1/(-0.5), 1/(-2)]$
$\quad\quad\quad = [\min(-1/(-2), -1/(-0.5), 1/(-2), 1/(-0.5))$
$\quad\quad\quad\quad \max(-1/(-2), -1/(-0.5), 1/(-2), 1/(-0.5))]$
$\quad\quad\quad = [\min(0.5, 2, -0.5, -2), \max(0.5, 2, -0.5, -2)]$
$\quad\quad\quad = [-2, 2]$

$[4, 10] / [1, 2] = [[4, 10] \cdot [1/2, 1/1]$
$\quad\quad\quad = [\min(4/1, 4/2, 10/1, 10/2), \max(4/1, 4/2, 10/1, 10/2)]$
$\quad\quad\quad = [\min(4, 2, 10, 5), \max(4, 2, 10, 5)]$
$\quad\quad\quad = [2, 10].$

8.3   ARITHMETIC OPERATIONS ON FUZZY NUMBERS

It goes without saying that in order for fuzzy numbers to be of practical use, we must be able to extend the usual operations on numbers, such as addition (+), subtraction (−), multiplication (·), and division (/), to fuzzy numbers. We know that fuzzy numbers are uniquely represented by their α-cuts and these are all closed intervals of real numbers. We also know how to apply the four basic operations to closed intervals. Hence, we can combine these insights to formulate the arithmetic operations on fuzzy numbers.

First, let us examine a special case. If we have two fuzzy numbers with triangular- or trapezoidal-shape membership functions, then we may perform the operations of addition and subtraction in a simple way—by adding or subtracting only the intervals corresponding to the cores and bases of the given triangles or trapezoids using rules of interval addition or subtraction. The resulting core and base characterize uniquely the resulting fuzzy number, which preserves the triangular or trapezoidal shape.

Suppose that we want to perform these two operations on the triangular-shape fuzzy numbers A and B specified in Fig. 8.8, which may represent (in some context) the concepts of *approximately* 2 and *approximately* 4, respectively.

In performing addition, the first step is to add the apexes (which are cores in this case) of the two numbers, $2 + 4 = 6$. This will then become the apex of the resulting fuzzy sum. The sum of the bases, viewed here as closed intervals, is $[4, 8] = [1, 3] + [3, 5]$. The resulting fuzzy number is shown in Fig. 8.8c. Subtraction, $A - B$, is performed in a similar way; its result is shown in Fig. 8.8d.

The simple procedures described for addition and subtraction are not applicable to multiplication and division. The reason is that the triangular or trapezoidal shapes of fuzzy numbers are not preserved under these operations.

One way to formulate any of the four basic arithmetic operations on arbitrary fuzzy numbers is to represent the numbers by their α-cuts and employ interval arithmetic to the α-cuts. To explain how this is done, consider arbitrary fuzzy numbers A and B, and let $*$ denote any of the four interval arithmetic operations. Then, for each $\alpha \in (0, 1]$, the α-cut of $A * B$ is defined in terms of the α-cuts of A and B by the formula

$$^{\alpha}(A * B) = {}^{\alpha}A * {}^{\alpha}B \qquad (8.2)$$

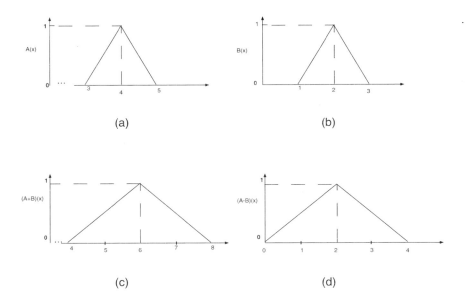

Figure 8.8   Simplified addition and subtraction of triangular fuzzy numbers.

which is not applicable when $*$ is division and $0 \in {}^{\alpha}B$ for any $\alpha \in (0, 1]$. Once the α-cuts ${}^{\alpha}(A * B)$ are determined, the resulting fuzzy number $A * B$ is readily expressed as

$$A * B = \bigcup_{\alpha \in [0,1]} {}^{\alpha}(A * B) \cdot \alpha$$

as explained in Sec. 5.2.

To illustrate the described procedure of performing arithmetic operations on fuzzy numbers, consider the triangular-shaped fuzzy numbers A and B shown in Fig. 8.9. These fuzzy numbers are expressed by the formulas

$$A(x) = \begin{cases} 0 & \text{for } x < -1 \text{ and } x > 3 \\ (x+1)/2 & \text{for } -1 \leq x \leq 1 \\ (3-x)/2 & \text{for } 1 \leq x \leq 3 \end{cases}$$

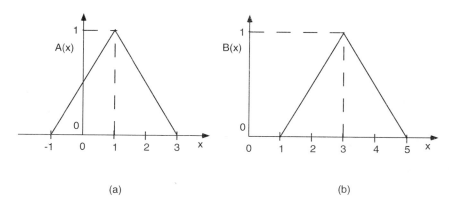

Figure 8.9 Fuzzy numbers used for illustrating arithmetic operations on fuzzy numbers.

$$B(\text{x}) = \begin{cases} 0 & \text{for } x < 1 \text{ and } x > 5 \\ (x-1)/2 & \text{for } 1 \le x \le 3 \\ (5-x)/2 & \text{for } 3 \le x \le 5 \end{cases}$$

For any given $\alpha \in (0, 1]$, the α-cuts of A and B can be uniquely determined from these formulas. To explain how this is done, let us introduce the notation

$$^{\alpha}A = [^{\alpha}a_1, {}^{\alpha}a_2]$$
$$^{\alpha}B = [^{\alpha}b_1, {}^{\alpha}b_2]$$

which is illustrated in Fig. 8.10. Then, clearly

$$A\,({}^{\alpha}a_1) = ({}^{\alpha}a_1 + 1)/2 = \alpha$$
$$A\,({}^{\alpha}a_2) = (3 - {}^{\alpha}a_2)/2 = \alpha$$

From these equations, we obtain

$$^{\alpha}a_1 = 2\alpha - 1$$
$$^{\alpha}a_2 = 3 - 2\alpha$$

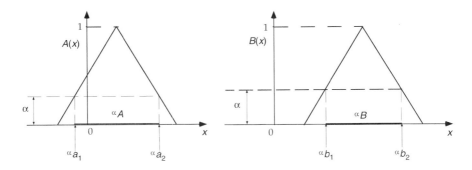

Figure 8.10   Construction of α-cuts of fuzzy numbers.

Hence,

$$^{\alpha}A = [2\alpha - 1,\, 3 - 2\alpha]$$

Similarly for fuzzy number B we have the equations

$$B\,(^{\alpha}b_1) = (^{\alpha}b_1 - 1)/2 = \alpha$$
$$B\,(^{\alpha}b_2) = (5 - {^{\alpha}b_2})/2 = \alpha$$

from which we obtain

$$^{\alpha}b_1 = 2\alpha + 1$$
$$^{\alpha}b_2 = 5 - 2\alpha$$

That is,

$$^{\alpha}B = [2\alpha + 1,\, 5 - 2\alpha]$$

Observe that we obtain general formulas for the closed inter-vals that characterize all α-cuts of the given fuzzy numbers A and B:

$$^{\alpha}A = [2\alpha - 1,\, 3 - 2\alpha]$$
$$^{\alpha}B = [2\alpha + 1,\, 5 - 2\alpha]$$

Substituting these intervals for $^{\alpha}A$ and $^{\alpha}B$ to Eq. (8.2), we have

$$^{\alpha}(A * B) = [2\alpha - 1,\, 3 - 2\alpha] * [2\alpha + 1,\, 5 - 2\alpha] \tag{8.3}$$

for the given fuzzy numbers A and B. When we interpret the general operation $*$ in Eq. (8.3) as the addition, the equation becomes specific:

$$^\alpha(A + B) = [4\alpha, 8 - 4\alpha]$$

which is the α-cut representation of the sum of the given fuzzy numbers A and B.

Since $\alpha \in (0, 1]$, the range of the left endpoint of the interval is $(0, 4]$ and the range of its right endpoint is $[4, 8)$. This means that

$$4\alpha = x \qquad \text{when } x \in (0, 4]$$

and

$$8 - 4\alpha = x \qquad \text{when } x \in [4, 8)$$

Solving these equations for α, we obtain

$$\alpha = x/4 = (A + B)\,(x) \qquad \text{when } x \in (0, 4]$$
$$\alpha = (8 - x)/4 = (A + B)\,(x) \qquad \text{when } x \in [4, 8)$$

That is, the membership function $A + B$ corresponding to the α-cut representation is expressed by the formula

$$(A + B)(x) = \begin{cases} 0 & \text{for } x < 0 \text{ and } x > 8 \\ x/4 & \text{for } 0 \le x \le 4 \\ (8-x)/4 & \text{for } 4 \le x \le 8 \end{cases}$$

Its graph is shown in Fig. 8.11a. Observe that we would obtain the same result by the simplified method explained at the beginning of this section.

Each of the three arithmetic operations on A and B can be determined in a similar way. For the difference $A - B$, we have

$$^\alpha(A - B) = [2\alpha - 1, 3 - 2\alpha] - [2\alpha + 1, 5 - 2\alpha] = [4\alpha - 6, 2 - 4\alpha]$$

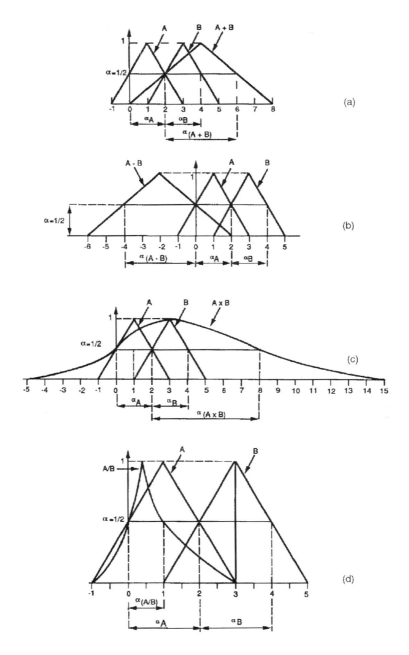

Figure 8.11   Arithmetic operations on the fuzzy numbers A and B specified in Figure 8.9.

and the corresponding membership function is

$$(A - B)(x) = \begin{cases} 0 & \text{for } x < -6 \text{ and } x > 2 \\ (x + 6)/4 & \text{for } -6 \leq x \leq -2 \\ (2-x)/4 & \text{for } -2 \leq x \leq 2 \end{cases}$$

The graph of $A - B$ is shown in Fig. 8.11b. Again, the same result would be obtained by the simplified method for triangular or trapezoidal fuzzy numbers.

To calculate the product $A \cdot B$, we again calculate the α-cut representation of $A \cdot B$ first:

$$^{\alpha}(A \cdot B) = [2\alpha - 1, 3 - 2\alpha] \cdot [2\alpha + 1, 5 - 2\alpha]$$

Applying the rule of interval multiplication, we find that the expression for the α-cuts are distinct in the intervals $\alpha \in (0, 0.5]$ and $\alpha \in (0.5, 1]$:

$$^{a}(A \cdot B) = \begin{cases} [-4\alpha^2 + 12\alpha - 5, 4\alpha^2 - 16\alpha + 15] & \text{for } \alpha \in (0,0.5] \\ [4\alpha^2 - 1, 4\alpha^2 - 16\alpha + 15] & \text{for } \alpha \in (0.5,1] \end{cases}$$

The resulting membership function $A \cdot B$, whose graph is shown in Fig. 8.11c, is defined by the formula

$$(A\ B)(x) = \begin{cases} 0 & \text{for } x < -5 \text{ and } x > 15 \\ [3 - (4 - x)^{1/2}]/2 & \text{for } -5 \leq x < 0 \\ (1 + x)^{1/2}/2 & \text{for } 0 \leq x < 3 \\ [4 - (1 + x)^{1/2}]/2 & \text{for } 3 \leq x \leq 15 \end{cases}$$

We can see that fuzzy multiplication does not preserve the triangular shape of the given fuzzy numbers. This means that the simplified method is not applicable to fuzzy multiplication.

To complete our example let us examine A/B. Using basically the same procedure described for the previous three operations, we obtain the α-cut representation

$$\alpha_{(A/B)} = \begin{cases} [(2\alpha - 1)/(2\alpha + 1),(3 - 2\alpha)/(2\alpha + 1)] & \text{for } \alpha \in (0, 0.5] \\ [(2\alpha - 1)/(5-2\alpha),(3 - 2\alpha)/(2\alpha + 1)] & \text{for } \alpha \in (0.5, 1] \end{cases}$$

and the membership function

$$(A/B)(x) = \begin{cases} 0 & \text{for } x < -1 \text{ and } x > 3 \\ (x + 1)/(2 - 2x) & \text{for } -1 \leq x < 0 \\ (5x + 1)/(2x + 2) & \text{for } 0 \leq x < 1/3 \\ (3 - x)/(2x + 2) & \text{for } 1/3 \leq x \leq 3 \end{cases}$$

This membership function is shown in Fig. 8.11d. We can see that fuzzy division does not preserve the triangular shape of the given numbers.

EXERCISES

8.1 Find examples of some fuzzy numbers encountered in daily life.

8.2 Consider two persons. The age of one person is estimated between 25 and 35. The age of the other person is estimated between 15 and 20. What is the estimated sum of the ages of the two persons? How much older is the first person?

8.3 Total annual salaries paid by a company are estimated between $1 and $3 million. The number of employees of the company is estimated between 50 and 100. What is the estimated average income of an employee of the company?

8.4 Let $[a, b]$ be an interval. Calculate the following:
  (a) $[a, b] + [a, b]$
  (b) $[a, b] - [a, b]$
  (c) $[a, b] \times [a, b]$
  (d) $[a, b] / [a, b]$, when $0 \notin [a, b]$

8.5 Let A and B be two fuzzy numbers whose membership functions are specified in Exercise 4.6. Calculate $A + B, A - B, A \cdot B$ and A / B.

8.6 Using the extension principle, discussed in Chapter 4, we can generalize the usual arithmetic of real numbers to the arithmetic of fuzzy numbers. For any binary operation $* \in \{+, -, \cdot ,/\}$, and any two fuzzy numbers A and B, we define $A*B$ by the formula

$(A*B)(x) = \sup\{\min(A(y), B(z))|$ for any y and z such that $y*z = x\}$

Show that $^{\alpha}(A*B) = {}^{\alpha}A*{}^{\alpha}B$ for any $\alpha \in [0,1]$. Does this mean that the arithmetic of fuzzy numbers is cutworthy?

8.7 Derive the formulas for $A - B$, $A \cdot B$ and A / B given in Sec. 8.3 for fuzzy numbers defined in Fig. 8.9.

8.8 Let V and W be trapezoidal-shaped fuzzy numbers, each defined by the four real numbers a, b, c, d shown in Fig. 8.3: V is defined by 1, 2, 3, 4; W is defined by 1, 2, 2, 3. Draw graphs of V and W and calculate:

(a) $V + V$, $W + W$, $V + W$, and $W + V$

(b) $V - V$, $W - W$, $V - W$, and $W - V$

(c) $V \cdot V$, $W \cdot W$, $V \cdot W$, and $W \cdot V$

(d) V / V, W / W, V / W, and W / V

8.9 Repeat Exercise 8.8 for another pair of trapezoidal-shaped fuzzy numbers: V is defined by -4, 3, 3, 1; W is defined by 1, 3, 4, 5.

9

FUZZY LOGIC

9.1 INTRODUCTION

The term "fuzzy logic" has been used in the literature in two different senses. In a broad sense, fuzzy logic is viewed as a system of concepts, principles, and methods for dealing with modes of reasoning that are approximate rather than exact. In an alternative, narrow sense, it is viewed as a generalization of the various multivalued logics, which have been studied in the area of symbolic logic since the beginning of this century. Our primary interest in this chapter is to examine fuzzy logic in the broad sense. However, to distinguish the two meanings of fuzzy logic, we also include an overview of the basic ideas of multivalued logics.

In its broad sense, fuzzy logic is an application area of fuzzy set theory. It utilizes concepts, principles, and methods developed within fuzzy set theory for formulating various forms of sound approximate reasoning. In order to utilize the apparatus of fuzzy set theory for approximate reasoning, it is necessary to establish a connection between degrees of membership in fuzzy sets and degrees of truth of fuzzy propositions.

Given a fuzzy set A, the membership degree $A(x)$ for any element x in the underlying universal set X may be interpreted as the degree of truth of the fuzzy proposition "x is a member of A." Conversely, given an arbitrary proposition "x is F", where x is from the set X and F is a fuzzy linguistic expression (such as *low*, *high*, *very far*, *extremely slow*, etc.), its degree of truth may be interpreted as the membership degree $A(x)$ by which a fuzzy set A characterized by the linguistic expression F is defined in a given context. Under this correspondence, operations of negation, conjunction, and disjunction on fuzzy propositions are defined in exactly the same way as the operations of complementation, intersection, and union on fuzzy sets, respectively. This correspondence is also essential for developing additional concepts needed in fuzzy logic, such as fuzzy truth qualifiers, fuzzy quantifiers, fuzzy probabilities, and so forth.

9.2 MULTIVALUED LOGICS

As we recognized earlier, all propositions in classical logic are either completely true or completely false, so our inferences are constrained by the awareness of being forced to recognize only these two truth value alternatives. Throughout the history of Western logic, there have been attempts to expand the rigid framework of two-valued logic and to allow inferences to include propositions whose truth values might be partly true and partly false. For example, Aristotle argues in his work *On Interpretation* that propositions about future events are neither true nor false, since the events to which the propositions correspond have not yet occurred and have not yet provided a truth value to these propositions: Propositions about the future are potentially true or potentially false. Hence their truth values are undetermined before the event. Of course, sooner or later, the future becomes the present and then the propositions about the now-present event will acquire truth values. But many propositions with indeterminate truth values do not have their truth values resolved so easily.

It is now well understood that not only propositions about future events have problematic truth status. Especially in fields such as quantum mechanics, the truth values of certain propositions are also inherently indeterminate because of the fundamental limitations of measurement of subatomic phenomena. This is an insight given to us by the famous Heisenberg uncertainty principle. Accordingly, in order to deal with such propositions, we must consider building logical frameworks that take into account the uncertainty of truth values. Such alternative logics are called *multivalued logics*.

Three-Valued Logics

All multivalued logics begin by relaxing the true/false dichotomy of classical two-valued logic by allowing one or more additional truth values between these two extremes: These values may be called *indeterminate*. In the case of a *three-valued logic*, there is exactly one indeterminate truth value. Several three-valued logics, each with its own rationale, are now well established and they commonly denote truth, falsity, and indeterminacy by 1, 0, and 1/2, respectively.

The introduction of this new intermediate truth value naturally affects the truth-table definitions of the five connectives discussed in classical logic, but since the proponents of various three-valued logics rely on their intuitions about the meanings of complex propositions containing indeterminate truth values, they do not all agree on the three-valued definitions of the connectives. The only exception—depicted in Table 9.1—is the negation $\neg p$ of a proposition p, which is commonly defined as $1-p$.

TABLE 9.1   THREE-VALUED NEGATION

p	$\neg p$
0	1
1/2	1/2
1	0

The definitions of the other four connectives—$\wedge$, $\vee$, $\Rightarrow$, and $\Leftrightarrow$—differ from one three-valued logic to another. Five of the best-known three-valued logics, labeled by the names of the logicians who originated them, are characterized in Table 9.2 by the ways in which they define these four logic connectives in terms of the three truth values: We can see in this table that the usual, classical definitions of the five connectives in terms of the truth values 0 and 1 are preserved. However, they differ from the classical definitions and from each other in the way in which they treat the new truth value 1/2. This fact has some important consequences.

One of these consequences is that none of the three-valued logics introduced satisfies the *law of contradiction* ($p \wedge \neg p = 0$), the *law of the excluded middle* ($p \vee \neg p = 1$), and some other tautologies—proposi-

TABLE 9.2   CONNECTIVES OF SOME THREE-VALUED LOGICS

a	b	Lukasiewicz $\wedge$	$\vee$	$\Rightarrow$	$\Leftrightarrow$	Bochvar $\wedge$	$\vee$	$\Rightarrow$	$\Leftrightarrow$	Kleene $\wedge$	$\vee$	$\Rightarrow$	$\Leftrightarrow$	Heyting $\wedge$	$\vee$	$\Rightarrow$	$\Leftrightarrow$	Reichenbach $\wedge$	$\vee$	$\Rightarrow$	$\Leftrightarrow$
0	0	0	0	1	1	0	0	1	1	0	0	1	1	0	0	1	1	0	0	1	1
0	½	0	½	1	½	½	½	½	½	0	½	1	½	0	½	1	0	0	½	1	½
0	1	0	1	1	0	0	1	1	0	0	1	1	0	0	1	1	0	0	1	1	0
½	0	0	½	½	½	½	½	½	½	0	½	½	½	0	½	0	0	0	½	½	½
½	½	½	½	1	1	½	½	½	½	½	½	½	½	½	½	1	1	½	½	1	1
½	1	½	1	1	½	½	½	½	½	½	1	1	½	½	1	1	½	½	1	1	½
1	0	0	1	0	0	0	1	0	0	0	1	0	0	0	1	0	0	0	1	0	0
1	½	½	1	½	½	½	½	½	½	½	1	½	½	½	1	½	½	½	1	½	½
1	1	1	1	1	1	1	1	1	1	1	1	1	1	1	1	1	1	1	1	1	1

tions that are always completely true—of two-valued logic. For example, the Bochvar three-valued logic shown in Table 9.2 clearly does not produce any of the tautologies of two-valued logic, since each of its connectives produces the truth value 1/2 whenever at least one of the atomic propositions constituting the putative tautologies assumes the value 1/2. Hence, in the Bochvar logic no classical tautology would ever achieve the value TRUE (1) in any row of its truth table.

Because of results such as this, it is common in three-valued logic to extend the usual concept of the tautology to the broader concept of a *quasi-tautology*. This concept is broader because it accepts as legitimate even propositions with truth values less than 1: We say that a logic formula in a three-valued logic which does not assume the truth value 0 (FALSE), regardless of the truth values assigned to its constituent propositional variables is a quasi-tautology, a statement that is not necessarily true. Similarly, we say that a logic formula which does not assume the truth value 1 (TRUE) is a *quasi-contradiction*. To contrast the effects of each of two different three-valued logics on a classical tautology, consider Tables 9.3, 9.4, and 9.5, which represent the truth tables for one of the De Morgan laws, using the Bochvar, Lukasiewicz, and Kleene logics, respectively.

One shared characteristic of the Bochvar and Kleene truth tables is that the intermediate value 1/2 occurs in some rows under the main connective, the equivalence, in each table. Hence, the De Morgan law is not a classical tautology in this logic. They differ, however, in the exact rows in which the proposition is evaluated as 1/2: In row 2 of their respective truth tables, it becomes clear that Bochvar is more liberal than Kleene on the conditions under which the conjunction may be true, but more restrictive than Kleene on the truth conditions of the disjunction. By contrast, in the Lukasiewicz logic, all the values under

TABLE 9.3 BOCHVAR THREE-VALUED INTERPRETATION
OF DE MORGAN'S LAW

p	q	$\neg$	$(p$	$\wedge$	$q)$	$\Leftrightarrow$	$(\neg p$	$\vee$	$\neg q)$
0	0	1	0	0	0	1	1	1	1
0	1/2	1/2	0	1/2	1/2	1/2	1	1/2	1/2
0	1	1	0	0	1	1	1	1	0
1/2	0	1/2	1/2	1/2	0	1/2	1/2	1/2	1
1/2	1/2	1/2	1/2	1/2	1/2	1/2	1/2	1/2	1/2
1/2	1	1/2	1/2	1/2	1	1/2	1/2	1/2	0
1	0	1	1	0	0	1	0	1	1
1	1/2	1/2	1	1/2	1/2	1/2	0	1/2	1/2
1	1	0	1	1	1	1	0	0	0

TABLE 9.4 LUKASIEWICZ THREE-VALUED INTERPRETATION
OF DE MORGAN'S LAW

p	q	$\neg$	$(p$	$\wedge$	$q)$	$\Leftrightarrow$	$(\neg p$	$\vee$	$\neg q)$
0	0	1	0	0	0	1	1	1	1
0	1/2	1	0	0	1/2	1	1	1	1/2
0	1	1	0	0	1	1	1	1	0
1/2	0	1	1/2	0	0	1	1/2	1	1
1/2	1/2	1/2	1/2	1/2	1/2	1	1/2	1/2	1/2
1/2	1	1/2	1/2	1/2	1	1	1/2	1/2	0
1	0	1	1	0	0	1	0	1	1
1	1/2	1/2	1	1/2	1/2	1	0	1/2	1/2
1	1	0	1	1	1	1	0	0	0

the equivalence are 1, and so this logic evaluates the De Morgan law in the same way as does two-valued logic. Lukasiewicz—like Kleene—is stricter than Bochvar on the conditions under which he is willing to count a conjunction as true. Like Kleene, he is more liberal than Bochvar on the truth conditions of disjunction. However, Lukasiewicz departs from Bochvar and Kleene by allowing the equivalence to be true, even when both elements just have the truth value 1/2.

Not only does the introduction of an intermediate truth value have an effect on the concepts of tautology and contradiction, but it also entails a change in how we think of rules of inference. Recall that

TABLE 9.5   KLEENE THREE-VALUED INTERPRETATION
OF DE MORGAN'S LAW

p	q	$\neg$	$(p$	$\wedge$	$q)$	$\Leftrightarrow$	$(\neg p$	$\vee$	$\neg q)$
0	0	1	0	0	0	1	1	1	1
0	1/2	1	0	0	1/2	1	1	1	1/2
0	1	1	0	0	1	1	1	1	0
1/2	0	1	1/2	0	0	1	1/2	1	1
1/2	1/2	1/2	1/2	1/2	1/2	1/2	1/2	1/2	1/2
1/2	1	1/2	1/2	1/2	1	1/2	1/2	1/2	0
1	0	1	1	0	0	1	0	1	1
1	1/2	1/2	1	1/2	1/2	1/2	0	1/2	1/2
1	1	0	1	1	1	1	0	0	0

every deductive rule of inference may be stated in a proposition which
is a tautology. Modus ponens, for example, may be expressed as the
proposition

$$[(p \Rightarrow q) \wedge p] \Rightarrow q$$

and is a tautology on a classical truth table. However, it is only a
quasi-tautology in the three-valued logics, as illustrated for the
Lukasiewicz logic in Table 9.6. Reasoning from the information provided in this table, we can say that because the proposition describing
modus ponens is not a tautology, the inference rule is not strictly valid.
We may expect, then, a broadening of the concept of validity along
with a broadening of the concept of truth.

TABLE 9.6   LUKASIEWICZ INTERPRETATION OF MODUS PONENS

p	q	$[(p$	$\Rightarrow$	$q)$	$\wedge$	$p]$	$\Rightarrow$	q
0	0	0	1	0	0	0	1	0
0	1/2	0	1	1/2	0	0	1	1/2
0	1	0	1	1	0	0	1	1
1/2	0	1/2	1/2	0	1/2	1/2	1/2	0
1/2	1/2	1/2	1	1/2	1/2	1/2	1	1/2
1/2	1	1/2	1	1	1/2	1/2	1	1
1	0	1	0	0	0	1	1	0
1	1/2	1	1/2	1/2	1/2	1	1	1/2
1	1	1	1	1	1	1	1	1

A final question we might ask is: What is the justification for deciding either to interpret a connective liberally or to interpret it restrictively? One answer is that the interpretation of logical connectives is intended to capture some basic intuition about our concept of a true proposition: We expect, for example, that no implication may be true if it has a true antecedent, but a false consequent (otherwise the entire concept of correct reasoning makes no sense). Thus, the interpretation of propositions in terms of three truth values should also follow some basic intuition about partial truth derived from considered applications; but, as we have seen, we do not all have the same applications in mind and, hence, the same intuitions. This is one reason that the truth tables for our sample tautology differed from each other.

n-Valued Logics

Once the various three-valued logics were accepted as meaningful and useful, the focus of interest fell on the idea that there could be "many-valued" logics as well, with their own interpretations and truth conditions. Such generalized logics were developed in the 1930s, a decade in which symbolic logic made a great number of advances.

Many-valued logics are usually referred to as *n-valued logics,* where n is the number of truth values a proposition may have in some particular logic. The idea behind n-valued logics is that, once we are allowed to think of a proposition p as having a truth value of 1/2—halfway between completely true and completely false—then we can think of propositions that are mostly true and only a little false. This truth value of a proposition p might be represented by the number 3/4. Similarly, a proposition p that is mostly false might have a truth value of only 1/4. Further, propositions that are hardly false could be assigned a truth value of 7/8, for example. Thus, we can see that an n-valued logic is quite useful for representing a wide range of somewhat true and somewhat false propositions which classical logic insists on burdening with just one of two extreme truth values.

For any given n, the truth values in these generalized logics are usually labeled by rational numbers in the unit interval [0, 1] that are obtained by evenly dividing the interval into $n - 1$ subintervals and taking their endpoints as the truth values. These are obtained by dividing each of the n values 0, 1,..., $n-1$ by $n-1$. That is, the set of truth values of an n-valued logic, T_n, is defined as

$$T_n = \left\{ \frac{0}{n-1}, \frac{1}{n-1}, \frac{2}{n-1}, \ldots, \frac{n-2}{n-1}, \frac{n-1}{n-1} \right\}$$

$$\ldots$$

$$= \left\{ 0, \frac{1}{n-1}, \frac{2}{n-1}, \ldots, \frac{n-2}{n-1}, 1 \right\}$$

Thus, in a five-valued logic, the truth values are 0, 1/4, 1/2, 3/4, and 1.

These values, it is becoming obvious, can be interpreted as *degrees of truth*, and so multivalued logics may be seen as the precursors of fuzzy logic. To see the connection, we can take a look at an n-valued logic proposed by the great Polish logician Lukasiewicz for any $n \geq 2$ (i.e., n may go to infinity). This is a generalization of his three-valued logic whose truth table is shown on Table 9.2. He uses truth values in T_n and defines the behavior of the five logical connectives by the following equations:

$$\bar{p} = 1 - p$$

$$p \wedge q = \min(p, q)$$

$$p \vee q = \max(p, q) \qquad\qquad (9.1)$$

$$p \Rightarrow q = \min(1, 1 - p + q)$$

$$p \Leftrightarrow q = 1 - |p - q|$$

These definitions remind us of the definitions presented in Chapter 2 in the discussion of the truth-value definitions of the classical connectives; indeed, if we apply these equations to the case of $n = 2$, where $T_2 = \{0, 1\}$, we obtain the traditional truth tables.

When truth values are not restricted to rational numbers but include all real numbers in the unit interval [0, 1], we obtain an *infinite-valued logic*. This logic is different from the infinite-valued logic based on rational truth values in the set T_n for $n \rightarrow \infty$. Since all values in the continuum [0,1] are used, it is suggestive to call such a logic a *continuous logic*. For example, by allowing the truth values of propositional variables in Eq. (9.1) to be real numbers in [0, 1], we obtain Lukasiewicz continuous logic. This is a special instance of fuzzy logic in the narrow sense. It is special in the sense that its logic connectives are based on the standard operations of fuzzy complement, intersection, and union, and on the special definitions of implication and equivalence given in Eq. (9.1). Other varieties of fuzzy logic are based on alternative definitions of the connectives. Fuzzy logic in the narrow sense may thus be conceived as the class of all logics with truth values in [0, 1].

As already mentioned, we are primarily interested in this text in the broad meaning of fuzzy logic. In this sense, the focus is on reasoning with propositions involving imprecise concepts, which are rather typical in natural language. This reasoning is usually referred to as *approximate reasoning*.

The following inference is an example of approximate reasoning in natural language that cannot be adequately dealt with in classical logic:

> Old coins are usually rare collectibles
> Rare collectibles are expensive
> ―――――――――――――――――――――――
> ∴ Old coins are usually expensive

This is a meaningful deductive inference, but since some of the predicate terms—*old, rare, usually, expensive*—are not precise, classical logic does not recognize this inference as a valid form. It is the purpose of fuzzy logic in its broad sense to deal with inferences of this kind—inferences based on fuzzy propositions expressed in natural language. The linguistic expressions involved may contain fuzzy linguistic terms of several types, including

- *fuzzy predicates*, such as tall, young, small, medium, normal, expensive, near, intelligent, and the like
- *fuzzy truth values*, such as true, false, fairly true, or very true
- *fuzzy probabilities*, such as likely, unlikely, very likely, or highly unlikely
- *fuzzy quantifiers*, such as many, few, most, or almost all

All these linguistic terms are represented in each context by appropriate fuzzy sets.

In the rest of this chapter, we examine basic types of fuzzy propositions and, then, some rules of inference based on fuzzy propositions.

We begin in the next section with propositions that do not involve fuzzy quantifiers.

9.3 FUZZY PROPOSITIONS

The fundamental difference between classical propositions and fuzzy propositions is in the range of their truth values. While each classical proposition is required to be either just true or just false, the truth or falsity of fuzzy propositions is a matter of degree. Assuming— as we have throughout—that truth or falsity are expressed by the values 1 and 0, respectively, the degree of truth of each fuzzy proposition is expressed by a number in the unit interval [0, 1].

Take, for example, the fuzzy proposition

(a) Mount Washington is a dangerous mountain.

We can see that the term *dangerous mountain* is not precise. For one thing, there is no sharp boundary between a tall hill and a short mountain, and—for another—the concept *dangerous* is not precise. However, the focus of our discussion in this section is on propositions, not objects in the world. That is, in this section we are not interested primarily in Mount Washington, for example, or in the question of its membership in the sets of mountains or dangerous things. We are interested in these matters only secondarily. Mainly, we are concerned with the assessment of the truth value of a fuzzy proposition: a proposition whose fuzziness may arise from a combination of different linguistic components. Thus, we are really considering the degree of truth of the proposition.

(b) 'Mount Washington is a dangerous mountain' *is true*.

There is an important difference between the propositional forms of (a) and (b): The former is a proposition in which a property is attributed to some object or event in the real world, so, it is said to be in the *object language*. The latter form is not a sentence about objects or events in the real world, but about *another* proposition. It says that some proposition has a property, and that the property is a degree of truth. Since proposition (b) is about a proposition, it is not in the object language but in the *metalanguaɡe*. We place single quotes around the proposition we are talking about to indicate that we are attributing a property to it.

We now introduce fuzzy propositions without quantifiers. In a crude way, it is useful to distinguish the following four types of fuzzy propositions:

- unconditional and unqualified propositions
- unconditional and qualified propositions
- conditional and unqualified propositions
- conditional and qualified propositions

Unconditional and Unqualified Propositions

Unconditional propositions are assertions that are not in conditional if-then form. Unqualified propositions are asserted to be simply true. Their truth values are not qualified by any modifying expressions. The symbolization of a fuzzy proposition of this type, p, is usually expressed by the propositional form

$$p: X \text{ is } A \tag{9.2}$$

We assume that X is, in general, a variable, such as temperature, that takes specific values x, such as 35°C, from a universal set of possible values (temperature readings); A is some property, or predicate, attributed to the variable. This property A is represented by an appropriate fuzzy set. In the case of temperature, A may be represented by the fuzzy set defining the predicate *high*. Remember that, since p is just a propositional pattern, it is a *propositional function* having no particular truth value. When its variable is replaced by a specific instance, we then obtain a proposition. The following proposition illustrates one instance of the symbolization:

The temperature 35°C is high.

The value 35°C is the value of the variable—temperature; and "is high" is the fuzzy predicate attributing a level to the reading of 35°C.

As was indicated earlier, when we assert such a fuzzy proposition we are actually saying that the *proposition p is true*, even though we do not explicitly do so in ordinary discourse. Hence, in ordinary discourse the propositional form (9.2) is tacitly assumed to stand for the form

$$p: \text{`}X \text{ is } A\text{'} \text{ is true} \qquad (9.3)$$

We need to explain now how the degree of truth of p is determined for a particular instance (value) x of X.

Given a particular value x of X, say 35°C, this x belongs to A with membership grade $A(x)$. This membership grade is then interpreted as the degree of truth, $T(p_x)$, of proposition

$$p_x: X = x \text{ is } A$$

That is,

$$T(p_x) = A(x) \qquad (9.4)$$

for each given value x of variable X in proposition p_x. In English, this symbolic expression says that the truth value of the proposition p_x is equal to the membership degree $A(x)$. The role of function T is to provide us with a bridge between fuzzy sets and fuzzy propositions. It is an identity function that assigns to any number in [0, 1] (representing the membership degree of x in A) the same number in [0, 1] (representing the degree of truth of proposition p_x).

As an example, let X be the relative humidity (measured in %) at some particular place on the Earth, and let the property of *high humidity* be expressed by membership function H shown in Fig. 9.1a. Then, we can apply the propositional form

$$p: X \text{ is } H$$

to various specific measurements x of X. For each $x \in [0, 100]$, the degree of truth, $T(p_x)$, of the fuzzy proposition

$$p_x: X = x \text{ is } H$$

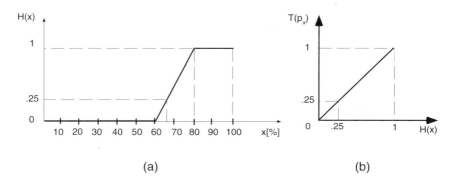

Figure 9.1   The degree of truth of the proposition "Humidity is high" when the measured humidity is 65%.

is defined by the identity function T from $[0, 1]$ to $[0, 1]$, which is shown in Fig. 9.1b. For example, if $x = 65$, then $H(65) = 0.25$, and we obtain

$$T(p_{65}) = H(65) = 0.25$$

That is, according the conception of *high humidity* expressed in Fig. 9.1a, the proposition

$$p_{65}: \text{Humidity of 65\% is high}$$

is true to the degree of 0.25.

Unconditional and Qualified Propositions

These are propositions making the unconditional assertion that another proposition has a qualified truth value. Propositions of this type are characterized by the form

$$p: \text{'}X \text{ is } A\text{' is } S \tag{9.5}$$

where p, X, and A have the same meaning as in (9.3) and S is a *fuzzy truth qualifier*, a linguistic expression that adds a modifier to the claim

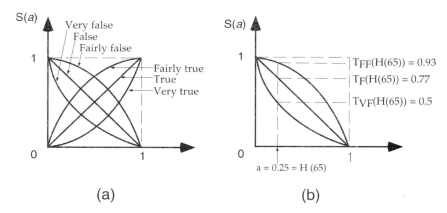

Figure 9.2   Illustration of the role of truth qualification.

of simple truth. Accordingly, we say that propositions of this form are *truth qualified*.

Examples of truth qualifiers are linguistic expressions such as *very true*, *fairly true*, *false*, *very false*, or *fairly false*. Each truth qualifier is characterized by a function from [0, 1] to [0, 1]. Possible functions attempting to capture the meaning of the mentioned linguistic expressions are shown in Fig. 9.2a. Also included is the identity function for the linguistic term *true*, which is employed in fuzzy propositions referred to as unqualified. This shows that unqualified fuzzy propositions are, in fact, special truth-qualified propositions in which the truth qualifier is assumed to be *true*.

In general, the degree of truth, $T_s(p_x)$, of the truth-qualified proposition

$$p_x: \text{'}\mathcal{X} = x \text{ is } A\text{' is } S$$

is determined for each $x \in X$ by the equation

$$T_s(p_x) = S(A(x)) \tag{9.6}$$

As an example, consider again the concept of *high humidity* defined by the membership function H in Fig. 9.1a and assume that the actual value of humidity is 65%. Then, $H(65) = 0.25$, and for any given truth qualifier S, we obtain the degree of truth of the proposition

$$p_{65}: \text{'Humidity of 65\% is high' is } S$$

by the formula

$$T_s(p_{65}) = S(0.25)$$

This is shown in Fig. 9.2b for truth qualifiers *fairly false (FF)*, *false (F)*, and *very false (VF)*.

Fuzzy logic can also handle propositions that include probability qualifications in addition to the various truth qualification. An example of such a proposition is: " 'The probability that the humidity (at some specified place and time) is high is very likely' is fairly true." Procedures for dealing with these propositions are more complicated and we do not deem it essential to cover them in this introductory book.

Conditional and Unqualified Propositions

Conditional and unqualified propositions are propositions in the "if-then" form that are assumed to be simply true. They are *fuzzy implications* and contain simple fuzzy propositions as antecedent and consequent. Propositions of this type are expressed by the form

$$p: \text{If } X \text{ is } A, \text{ then } Y \text{ is } B \qquad (9.7)$$

where X and Y are variables that take values x and y from sets X and Y, respectively, and A, B are relevant predicates represented by fuzzy sets. Of course, as before, it is assumed that (9.7) stands actually for the form

$$p: \text{'If } X \text{ is } A, \text{ then } Y \text{ is } B \text{' is true} \qquad (9.8)$$

Specific instances of 'X is A' and 'Y is B' are represented as $A(x)$ and $B(y)$, respectively:

$$p_{x,y}: \text{'If } A(x), \text{ then } B(y) \text{' is true} \qquad (9.9)$$

An example of such a conditional proposition is

'If Tina is young, then John is old' is true.

In analogy with classical logic, we can symbolize the relation between the two components as a fuzzy implication

$$A(x) \Rightarrow B(y)$$

However, contrary to the uniqueness of implication in classical logic, there is a class of functions on $[0, 1]^2$ that qualify as fuzzy implications. A particular function must be chosen from the class in the context of each application. The most common fuzzy implication is the Lukasiewicz implication introduced in Sec. 9.2. For the sake of simplicity, we restrict ourself in this text to this particular fuzzy implication. For each $x \in X$ and each $y \in Y$, the Lukasiewicz implication I determines the degree of truth of the conditional proposition (9.9) by the formula

$$T(p_{x,\, y}) = I \, [A(x), B(x)] = \min[1, 1 - A(x) + B(x)] \qquad (9.10)$$

As an example, consider the conditional and unqualified fuzzy proposition of the form

p: If a textbook is large, then it is expensive.

where the concepts of a large textbook and an expensive textbook are expressed, respectively, by the fuzzy sets L and E in Fig. 9.3. For each particular textbook, we obtain a fuzzy proposition $p_{x,\, y}$, which has some specific degree of truth. This degree is calculated as follows:

(a) Given the size of the textbook, x, defined by its number of pages, we determine the value of $L(x)$, which characterizes the degree of compatibility of the size with the concept of large textbooks;

(b) given the price of the textbook, y, we determine the value $E(y)$, which characterizes the degree of compatibility of the price with the concept of expensive textbooks;

(c) we calculate the degree of truth, $T(p_{x,\, y})$, of proposition $p_{x,\, y}$ by applying the Lukasiewicz implication (or, more generally, the

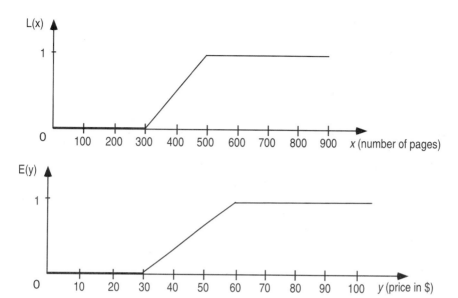

Figure 9.3   Fuzzy sets involved in conditional fuzzy proposition of the proposition "If a textbook is large, then it is expensive."

chosen fuzzy implication) to the determined values of $L(x)$ and $E(y)$:

$$T(p_{x, y}) = \min [1, 1 - L(x) + E(y)]$$

For a textbook that has 600 pages and costs \$45, we have $L(600) = 1$ and $E(45) = 0.5$. Hence, the degree of truth of the conditional proposition for this textbook is

$$T(p_{600, 45}) = \min [1, 1 - 1 + 0.5] = 0.5$$

For another textbook, which has 450 pages and costs \$42, we have $L(450) = 0.75$ and $E(42) = 0.4$. Hence,

$$T(p_{450, 42}) = \min [1, 1 - 0.75 + 0.4] = 0.65$$

For yet another textbook, which has 574 pages and costs \$60, we have $L(574) = 1$, $E(60) = 1$, and $T(p_{574, 60}) = 1$. Hence, the conditional proposition for this textbook is true.

Conditional and Qualified Propositions

Fuzzy propositions of this type are characterized by the propositional form

$$p: \text{'if } X \text{ is } A, \text{ then } B \text{ is } y' \text{ is } S$$

where—as before—the symbol S stands for a truth qualifier. Since methods introduced for the other types of propositions can be applied to deal with propositions of this type, we do not deem it necessary to discuss them further. To determine the degree of truth $T_s(p_{x,y})$ of a particular proposition of this type, we first determine the degree of truth $T(p_{x,y})$ of the associated unqualified proposition and then apply the truth qualifier S to it. That is,

$$T_s(p_{x,y}) = S[T(p_{x,y})] \tag{9.11}$$

Using, for example, the concepts of a large textbook and an expensive textbook as defined in Fig. 9.3, we can determine by (9.11) the degree of truth of any fuzzy proposition of the form

$$p: \text{'If a textbook is large, then it is expensive' is very true}$$

When applying this propositional form to the three textbooks example considered previously, we obtain fuzzy propositions whose degrees of truth are shown in Fig. 9.4.

9.4 *FUZZY QUANTIFIERS*

The two quantifiers of predicate logic—the universal quantifier *all* and the existential quantifier *there exists*—were introduced in Chapter 2. It is clear that the process of translating natural-language sentences into the language of quantifiers forces us to make a fairly crude distinction between whether we are talking about all the members of a set or only about some particular individuals.

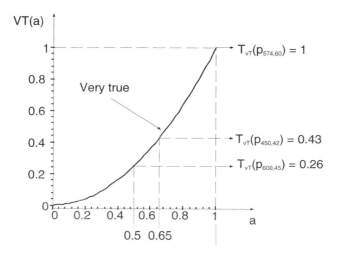

Figure 9.4 Degrees of truth of conditional and truth-qualified fuzzy propositions.

Of course when we are referring to all members of a set, there is no problem capturing the full meaning and truth of the natural-language sentence. For example, the sentence

All snakes are reptiles.

is uncontroversially translated as

For all x, if x is a snake, then x is a reptile, or

$$(\forall x)(Sx \Rightarrow Rx)$$

However, we face some uncomfortable choices when translating a less particular sentence. For example, the sentence

Some snakes are poisonous.

is translated in symbolic logic in the existential form

$$(\exists x)(Sx \wedge Px)$$

This sentence says that there exists at least one poisonous snake and expresses a minimal meaning of our original sentence: If the English sentence is true, then the symbolized one *must* be true.

Perhaps we are willing to strip our original English sentence of a certain amount of meaning because it attributes a characteristic to just a limited number of snakes. Further, classical logic is concerned more with displaying the *logical form* of propositions, not so much with displaying their *content*. But, if our sentence had been

Almost all snakes are poisonous.

then we would have to make a much larger sacrifice in meaning. Even though it makes a much more sweeping assertion, this new sentence would still have to be translated as

$$(\exists x)(Sx \wedge Px)$$

since we are not making a universal claim.

We can see, then, that being restricted to only two quantifiers has the result that we must give up a great deal of information about our subject in the effort to make the logical form of our assertion transparent. Fuzzy quantifiers are a tool for symbolizing quantified statements that minimizes the loss of information forced by the choice of quantifier.

Fuzzy propositions of any of the types introduced in Sec. 9.3 may be quantified by a suitable fuzzy quantifier. In general, fuzzy quantifiers are fuzzy numbers which take part in the various propositional forms and affect the degrees of truth of specific fuzzy propositions. Each fuzzy quantifier expresses an approximate number of elements or an approximate proportion of elements in a given universal set that are claimed to satisfy a given property.

Fuzzy quantifiers of one type, which are called *absolute quantifiers*, are expressed by fuzzy numbers defined on the set of real numbers or on the set of integers. They characterize linguistic terms such as *about a dozen*, *at most about 10*, *at least about 100*, and the like. Examples of fuzzy propositions with absolute quantifiers are

About 20 hotels are in close proximity to the center of the city.
At least about a dozen of applicants for the job have strong
    administrative experience.
At most about 10 students in the class are highly fluent in
    English.

Quantifiers of another type, which are called *relative quantifiers*, are expressed by fuzzy numbers defined on [0, 1]. They characterize linguistic terms such as *most, almost all, about half, about 20%,* and the like. Examples of quantifiers of this type are:

Almost all hotels in close proximity to the airport are expensive.
About half of the required textbooks in the program are
  expensive.
Most of the applicants for the job are young women.

Various procedures that are required to calculate degrees of truth of quantified fuzzy propositions, involving absolute quantifiers or relative quantifiers, are now well developed and described in the literature. However, their coverage is beyond the scope of this introductory text.

9.5 LINGUISTIC HEDGES

In this section, we examine certain linguistic modifiers that change the level of truth values or the strength of quantifiers. For example, we provide a framework for clarifying the common-sense insight that the truth value of a proposition such as "Tina is young" is different from that of the similar proposition "Tina is very young." The modifier *very* essentially focuses us on a meaning of *young* that is somewhat different from that of the unmodified concept of *young*. Similarly, adding the modifier *almost* to the universal quantifier changes the scope of our assertions.

The use of the fuzzy quantifier *almost all* is particularly interesting. Sometimes we use this phrase in a situation in which we actually would like to assert the mush stronger *all*, as in "all mountains are steep." However, being justified in asserting a universal proposition requires that we have complete evidence. It is not too difficult to make such propositions false. Of course, we usually do not have complete evidence, so—as a hedge against being wrong—we retreat to safer ground and assert a weaker proposition using the phrase *almost all*. The word *almost* serves as our hedge, which is why such expressions are called *linguistic hedges*.

Accordingly, linguistic hedges are special linguistic terms by which other linguistic terms are modified. Linguistic terms, such as *very, more* or *less, fairly,* or *extremely* are all examples of such hedges.

They can be used for modifying fuzzy predicates, fuzzy truth values, and fuzzy quantifiers. For example, a proposition of the form "x is young," which means that the proposition "x is young" is true, may be modified by the hedge *very* in any of the following three ways:

"x is very young" is true

"x is young" is very true,

"x is very young" is very true

In general, given a fuzzy proposition

$$p_x: \text{'}x \text{ is } A\text{'} \text{ is true}$$

and a linguistic hedge, H, we can construct a modified proposition. For example,

$$Hp_x: \text{"}x \text{ is } HA\text{"} \text{ is true}$$

where HA denotes the fuzzy predicate obtained by applying the hedge H to the given predicate A. Similarly, we can modify the truth qualifier of the proposition.

Any linguistic hedge, H, may be interpreted as a unary operation, h, on the unit interval $[0, 1]$. For example, when an explicit membership function for it is not available, the hedge *very* is often interpreted as the unary operation $h(a) = a^2$ ($a \in [0, 1]$), which means that *very* has the effect of squaring the truth value of the predicate it modifies. The hedge *fairly* is often interpreted as $h(a) = \sqrt{a}$ ($a \in [0, 1]$). Appropriately, we shall let unary operations that represent linguistic hedges be called *modifiers*. We must recognize, however, that this interpretation of the two operators here is a default interpretation; the actual operators that are chosen in the context of an application must be generated either from experimental data or from an analysis provided by a domain expert.

Given a fuzzy predicate A in the context of the universal set X and a modifier h representing a linguistic hedge H, the modified fuzzy predicate HA is determined for each $x \in X$ by the equation

$$HA(x) = h(A(x))$$

This just means that we may study the properties and effects of fuzzy hedges by studying the truth-changing properties of modifiers.

A *strong* modifier, such as *extremely* or *very*, strengthens the fuzzy predicate to which it is applied and, consequently, it reduces the truth value of the associated proposition. Thus the proposition "'Tina is very young' is very true" has a lower truth value than the proposition "'Tina is very young' is true." A *weak* modifier, to the contrary, weakens the predicate and, hence, the truth value of the proposition increases.

As an example, consider the following three fuzzy propositions (all of which are said to be simply true):

p_1: John is young
p_2: John is very young
p_3: John is fairly young

and let the linguistic hedges *fairly* and *very* be represented by the strong modifier a^2 and the weak modifier $\sqrt{a}$, respectively. Assume now that John is 26 years old and that, according to the adopted fuzzy set YOUNG representing the fuzzy predicate *young*, YOUNG(26) = 0.8. Then, VERY YOUNG(26) = 0.8^2 = 0.64 and FAIRLY YOUNG (26) = $\sqrt{0.8}$ = 0.89. Hence, $T(p_1)$ = 0.8, $T(p_2)$ = 0.64, $T(p_3)$ = 0.89. These values agree with our intuitions: The stronger assertion is less true, given the same data, and the weaker assertion is more true.

In representing modifiers as linguistic hedges, we must be careful to avoid various ambiguities of natural language. For example, the linguistic term *not very* may be viewed as the negation of the hedge *very*, but some authors argue that it should be regarded as a new hedge that is somewhat weaker than the simple hedge *very*.

9.6 APPROXIMATE REASONING

One of the aims of classical, symbolic logic is, as discussed in Chapter 2, to justify complex reasoning by reference to simple inference rules—such as *modus ponens* and *hypothetical syllogism*—that represent **forms** of inference. Let us recall the schema of the rule of modus ponens:

$$p \Rightarrow q$$
$$p$$
$$\overline{}$$
$$q$$

Another representation of this inference is the tautologous conditional that describes it:

$$[(p \Rightarrow q) \wedge p] \Rightarrow q$$

This inference works because we assume that the proposition p in the second premise is the same proposition p as in the antecedent of the first premise (and also that the two instances of q stand for exactly the same proposition). If they were not the same, we would not have a classically valid inference. The success of this important inference rule is due to the fact that it is based on already accepted rules about the truth of the conditional $p \Rightarrow q$. These rules, in turn, depend on accepting only 1 (TRUE) or 0 (FALSE) as truth values for the constituent, atomic sentences. We know that, in order to guarantee the truth of the conclusion of a deductive inference, we must establish the truth of the premises; thus, in order to carry out correct *deductive* reasoning, we must use a valid argument form and also employ true premises. Our task here is to apply similar principles to *approximate* reasoning.

Approximate reasoning is an important application area of fuzzy set theory. It is essential for modeling human common-sense reasoning. It is also important for expert systems in dealing with reasoning under a fuzzy environment.

In our daily lives, we almost always use common-sense reasoning, as exemplified by the following inference:

Rule:	If a book is large, then it is expensive	
Fact:	Book x is fairly large	(9.12)
Conclusion:	Book x is fairly expensive	

This kind of inference cannot be formulated in terms of classical two-valued logic for at least two reasons. One reason is that the concepts *large*, *fairly large*, and *expensive* are fuzzy concepts, which can not be adequately formulated as two-valued propositions. The other reason is that the classical modus ponens requires that the premise must match the antecedent of the if-then rule. In inference (9.12), this is not the case. Therefore, the classical modus ponens is not applicable.

Inference (9.12) is an example of the *generalized modus ponens* in approximate reasoning. Formally, it is represented by the scheme:

Rule:	If X is A, then Y is B	
Fact:	X is A'	(9.13)
Conclusion:	Y is B'	

where X, Y are variables taking values in the universal sets X and Y, respectively; A and A' are fuzzy sets on X; and B and B' are fuzzy sets on Y. Conclusion B' is calculated for any $y \in Y$ by the formula

$$B'(y) = \sup_{x \in X} \min (A'(x), I(A(x), B(y))) \qquad (9.14)$$

usually referred to as the *compositional rule of inference*, where I denotes an appropriate fuzzy implication.

We present the generalized modus ponens here only to illustrate the spirit of approximate reasoning. For a deeper study of approximate reasoning, which is currently a rapidly growing area, interested students should consult the graduate text mentioned in the Preface.

EXERCISES

9.1 Why is a three-valued logic needed? List some cases where we use three-valued logic in daily life.

9.2 Determine all possible truth values of the following logic formulas in the Lukasiewicz three-valued logic.

   (a) $p \vee \bar{p}$

   (b) $p \wedge \bar{p}$

   (c) $p \wedge (p \vee q)$

   (d) $p \vee (p \wedge q)$

9.3 For any $p, q \in T_n$, show that

   (a) $p \Rightarrow q = 1$ if and only if $p \leq q$;

   (b) $p \Leftrightarrow q = 1$ if and only if $p = q$

9.4 Read some articles in a newspaper and identify fuzzy propositions employed in these articles. For each identified fuzzy proposition, determine its type and its various components (fuzzy predicates, truth qualifiers, probability qualifiers, quantifiers, linguistic hedges).

9.5 Suppose we have a fuzzy proposition

      Today's humidity is high.

Using the fuzzy set defined in Fig. 9.1a, determine the truth value of the proposition for today's humidity, 50, 60, 70, 80, 90 and 100, respectively.

9.6 Suppose a 500-page textbook costs $60. Using the fuzzy sets specified in Fig. 9.3, calculate the truth values of the propositions:

   (a)   "If a textbook is large, then it is expensive" (is true).

(b) "If a textbook is not large, then it is not expensive" (is true).

(c) The propositions in (a) and (b) with the truth qualifier *very true*, where $very(a) = a^2$ for any $a \in [0,1]$.

(d) The propositions in (a) and (b) with the truth qualifier *fairly true*, where $fairly(a) = a^{1/2}$ for any $a \in [0,1]$.

9.7 Find some examples of approximate reasoning in daily life.

9.8 Consider the conditional fuzzy proposition "If a textbook is large, then it is expensive," in which the fuzzy predicates *large* and *expensive* are those defined in Fig. 9.3. Consider, in addition, the unconditional fuzzy proposition "a textbook has about 400 pages," in which the predicate 'about 400' is represented by triangular-shaped fuzzy number whose core and support are 400 and (300, 500), respectively. Using the generalized modus ponens, determine the approximate price of the book.

10

APPLICATIONS: A SURVEY

Fuzzy set theory has lately attracted considerable attention, primarily due to its surprisingly successful and highly visible applications. Although the purpose of this text is to introduce students to the basic elements of fuzzy set theory independent of applications, it is desirable to discuss, at least briefly, our current evidence of the great practical utility of the theory. This is the aim of this closing chapter of the text.

A short historical overview regarding applications of fuzzy set theory is followed by a survey of both well-established and prospective applications of the theory. The chapter is then concluded with a few illustrative examples of applications. Unlike previous chapters, this chapter contains some carefully chosen references to the literature. These are intended for students who wish to learn more about specific application areas of fuzzy set theory.

10.1 A HISTORICAL OVERVIEW

From the standpoint of applications, three rather natural phases in the development of fuzzy set theory can be recognized.

1. The period 1965–77, often referred to as the *academic phase*, is characterized by the development of fundamentals of fuzzy set theory and only initial speculation about prospective

applications of the theory. The outcome was a rather small number of publications of a predominantly theoretical nature by a small number of contributors, primarily from the academic community.

2. The period 1978–88, referred to as the *transformation phase*, is characterized not only by significant advances in fuzzy set theory, but also by some successful practical applications of the theory. The number of contributors, some of them from industry and business, increased rapidly. This resulted in a substantial increase in relevant publications, some of which discussed the various emerging applications. It is also significant that some important professional societies and journals devoted to fuzzy set theory and its applications were established during this period.

3. The current period, which began in 1989 and is often referred to as the *fuzzy boom*, is characterized by a rapid increase in successful industrial and business applications of fuzzy set theory which has resulted in very impressive revenues. Some major companies, initially in Japan, endorsed fuzzy set theory and committed resources to its further development. Major research centers devoted to applications of the theory were established. This has all been accompanied by a tremendous increase in the number of relevant publications, including several dedicated journals. At the same time, computer software and hardware designed for various applications of fuzzy set theory have become commercially available in increasing variety. In the early 1990s, fuzzy set theory became recognized as one of the key ingredients of the emerging area of *soft computing*. The aim of soft computing is to exploit, whenever possible, the tolerance for imprecision and uncertainty in order to achieve computational tractability, robustness, and low cost by methods that produce acceptable approximate solutions to complex problems which are often not precisely formulated. In soft computing, the main partners of fuzzy set theory are neural networks and genetic algorithms.

The development of fuzzy set theory, and particularly its applications, is well and thoroughly depicted in the book *Fuzzy Logic* by D. Mc Neill and P. Freiberger (Simon & Schuster, 1993), which is highly recommended as a supplementary reading to this text. The evolution of key ideas of fuzzy set theory is documented best in two books of collected papers by Lotfi A. Zadeh [6, 14]; he is not only the founder of the theory but has also been the principal contributor to its development.

10.2   ESTABLISHED APPLICATIONS

The most successful area of applications of fuzzy set theory by far has been *fuzzy control*. The nature of fuzzy controllers varies substantially according to the control problems they are supposed to solve. Control problems range from complex tasks, typical in robotics, which require a multitude of coordinated actions, to simple goals, such as maintaining a prescribed state of a single variable.

While classical controllers are based on mathematical models of the controlled processes, fuzzy controllers are based, by and large, on knowledge elicited from human operators. This knowledge is formulated in terms of a set of fuzzy control rules, each expressed by a conditional fuzzy proposition, such as,

IF the speed is *negative low* (with respect to the desired speed)
AND the change of speed is *negative high*
THEN the change of power should be *positive high*

where *speed* and *change of speed* are observed variables of the controlled process, while *change of power* is a variable representing the action of the controller. The linguistic terms *negative low, negative high, positive high*, and the like, are represented by appropriate fuzzy numbers.

A simple fuzzy controller is described in Sec. 10.4. By this example, we explain how knowledge expressed by a set of fuzzy control rules is utilized by fuzzy controllers for producing appropriate control actions. For a deeper study we recommend references [2,13].

The spectrum of applications of different complexities in which fuzzy controllers are already well established is enormous. At one end of the spectrum are simple fuzzy controllers employed routinely in many consumer products: washing machines, vacuum cleaners, electric shavers, dishwashers, rice cookers, video camcorders, cars (for anti-skid brakes, automatic transmissions, speed controls, and other functions), refrigerators, humidifiers, and air conditioners, to name just a few examples. More complex fuzzy controllers have been employed, for example, for controlling groups of elevators, trains of subway systems, traffic in cities, and various industrial processes. One of the most complex fuzzy controllers, which has been successfully tested and implemented, is capable of controlling a helicopter according to instructions in natural language which are communicated to the helicopter from ground via wireless transmitter. This achievement is considered, in general, beyond the capabilities of classical control theory. Fuzzy controllers may also be combined with classical controllers to achieve a high performance. They are also often combined with appropriate neural networks whose learning capabilities are utilized for making the controllers adaptive to varying conditions.

Another very successful application area of fuzzy set theory, closely following the success of fuzzy controllers, is the broad area of *decision making*. Fuzzy methods have been developed in virtually all branches of decision making, including multiobjective, multiperson, and multistage decision making. These methods are, in general, more realistic than their classical counterparts.

The literature on fuzzy decision making is abundant. An early book on fuzzy decision making [4] is still pedagogically the best comprehensive introduction to the subject, even though it is not fully up to date. A more recent book on fuzzy decision making [15], written as a textbook, is also highly recommended.

Somewhat connected with fuzzy decision making are applications of fuzzy set theory in *management*, *business*, and *operations research*. Fuzzy methods are now well developed for the various problem classes pertaining to these areas, such as optimization (linear programing, dynamic programming, etc.), scheduling, planning, and the like. One of the textbooks on fuzzy set theory [16] has an extensive coverage of these applications.

Fuzzy set theory has already established an important role in *computer science*, particularly in dealing with the various issues involving the storage and manipulation of information and knowledge in a manner compatible with human thinking. This includes *fuzzy databases, fuzzy information retrieval systems*, and *fuzzy expert systems*. The principal advantage of using fuzzy sets in these computer-based systems is the gained capability of representing and manipulating information and knowledge expressed in natural language. This capability makes these systems more flexible and realistic.

Literature on the use of fuzzy sets in database systems, information-retrieval systems, and expert systems is quite extensive. For an introduction to these areas, we recommend [10] for fuzzy database systems; [8] for fuzzy information-retrieval systems; and [3] for fuzzy expert systems.

The utility of fuzzy set theory is also well established in the problem area of *pattern recognition*, as well as the associated areas of *cluster analysis* and *image processing*. A common thrust of these problem areas is the search for structures in data. In cluster analysis, the aim is to classify data into categories (clusters) such that the similarity is high within each category and low between categories. In pattern recognition, the issue is to compare, in terms of relevant features, the categories identified in data with given prototypical categories. Image processing is associated with cluster analysis and pattern recognition in the sense that data are often given in terms of digital images.

The use of fuzzy set theory in pattern recognition and other classification problems was already anticipated in the 1960s, shortly after the emergence of the theory. This is quite understandable since most categories we commonly encounter and use do not have precise boundaries. From among the many publications that deal with the various issues of fuzzy classification we recommend [9], which is a broad summary of these issues.

It is fair to say that applications of fuzzy set theory are considerably more developed in engineering than in science. Among engineering applications of fuzzy set theory, the most visible by far are fuzzy controllers. This great visibility is responsible for equating the role of fuzzy set theory in engineering with fuzzy controllers, which has been quite common in popular literature. Such a portrayal is misleading since many successful engineering applications of fuzzy set theory have nothing in common with fuzzy controllers.

Engineering applications were first recognized in *civil engineering*. This is not surprising since each civil engineering project is, by and large, unique and, hence, the dependence on human judgment and approximate "rules of thumb" is substantially higher than in the other engineering disciplines. The use of fuzzy sets has been successful, for example, in assessing or evaluating existing constructions (bridges, highway pavements, buildings). Subjective assessments of individual components of the constructions and their structural importance can conveniently be expressed by appropriate fuzzy numbers. The overall evaluation of the construction is then expressed as the weighted average, which is calculated by fuzzy arithmetic.

Fuzzy set theory plays also a significant role in *engineering design*. The basic idea is that fuzzy sets allow the designer to describe the designed artifact as approximately as desired at early stages of the design process. This is important, as recognized by experienced designers, since any early design decisions restrict the set of available design alternatives. If good alternatives in a design problem are eliminated by these early decisions, they cannot be recovered in later stages of the design process. That is, every early decision that is wrong is costly. If the early decisions are required to be precise, then it is virtually impossible to determine how good or bad each decision is in terms of the subsequent step and, eventually, in terms of the final design. Experienced designers recognize this difficulty and attempt to begin the design process with a complete but approximate description of the desired artifact. As the design process advances from the formative stage to more detailed design and analysis, the degree of imprecision in describing the artifact is reduced. At the end of the design cycle, the imprecision is virtually eliminated, except for unavoidable tolerances resulting from imperfections in the manufacturing process. Fuzzy set theory is eminently suited to automate this process. The approximate description of input design parameters by appropriate fuzzy sets is employed to calculate the corresponding approximate characterization of relevant output design parameters. The latter are compared with given performance criteria, and information obtained by this comparison is then utilized to determine appropriate values of input parameters. When functional dependencies are numerical, the calculations involve fuzzy arithmetic. Otherwise, the extension principle must be employed.

Literature on engineering applications of fuzzy set theory, including civil, mechanical, electrical, chemical, and nuclear engineering, is very extensive and rapidly growing. For an overview, a recent textbook [11] is recommended, which is oriented to engineering applications.

Many applications of fuzzy set theory are now facilitated by an appreciable number of commercially available software packages of various types and capabilities. Moreover, specialized hardware for fuzzy computing has been available commercially since the early 1990s. It allows us to increase operational speed via parallel processing and this in turn extends the scope of applicability of fuzzy set theory.

10.3 PROSPECTIVE APPLICATIONS

This section is an overview of applications of fuzzy set theory which are less visible than those described in Sec. 10.2. Some of these applications are now fairly well developed but not yet fully accepted by the professional communities involved.

The utility of fuzzy set theory in *medicine* has been recognized since the mid-1970s. The main focus has been on the use of fuzzy sets in modeling the process of diagnosis of disease. This is not surprising since diseases are usually characterized by statements in natural language. For example, cirrhosis is characterized by the statement

Total proteins are *decreased*, α-globulins are *frequently decreased*, and γ-globulins are *increased*.

where the linguistic terms printed in italics are inherently vague. To represent these terms meaningfully requires the use of fuzzy sets. Moreover, fuzzy sets are essential for describing the appearance of symptoms in patients and for describing the relations between symptoms and diseases.

It is clear from these observations that applications of fuzzy set theory in medicine have great potential. Some of them are already described in the literature, including fuzzy expert systems designed to aid the physician in diagnosis within some specified category of diseases. By and large, however, the utility of fuzzy set theory has not yet been endorsed by the medical community.

Another area in which the role of fuzzy set theory has recently been recognized is *economics*. It is likely that mathematics based on fuzzy sets will make economic theories more realistic and increase their predictive capabilities. The first effort in this direction is described in a recent book by the French economist A. Billot [1]. The growing interest in utilizing fuzzy set theory in economics is also manifested by the emergence of the International Association for Fuzzy-Set Management and Economy (founded in 1994) and its journal *Fuzzy Economic Review* (commenced in 1995). Although this important direction has thus far been endorsed only by a small number of economists, it will likely affect economics in a profound way in due time.

One area that has a special relationship with fuzzy set theory is *psychology*. On the one hand, fuzzy set theory has a great potential in psychology; unfortunately, this potential has not yet been fully utilized. On the other hand, psychology is important for further development of fuzzy set theory. Psycholinguistics in particular is essential for studying the connection between the human use of linguistic terms in different contexts and the fuzzy sets and fuzzy operations that represent these terms. This connection plays a crucial role in clarifying the notion of context, which is inevitably involved in any application of fuzzy set theory. Although the interest of psychologists in fuzzy set theory has been visible since the mid-1980s, the relevant literature is too dispersed to see any general trends.

As already mentioned, fuzzy sets have been applied substantially more in engineering than in science. In *natural sciences* (physics, chemistry, biology, ecology), applications of fuzzy set theory are particularly rare. Nevertheless, there is some evidence, particularly in *chemistry* [12], that the utility of fuzzy sets in these areas is beginning to be recognized. Thus far, the use of fuzzy sets in natural sciences has been explored in *mathematical chemistry, quantum physics, nonequilibrium thermodynamics, ecological modeling*, and *biological classification*.

A few other areas in which fuzzy sets have already been applied to various degrees should be at least mentioned. They include *reliability analysis of large-scale software systems, risk and hazard analysis, financial analysis, image and speech understanding, transportation, earthquake studies*, and *robotics*. A popular book by Kosko [7] is recommended as an interesting supplementary reading about existing as well as potential applications of fuzzy set theory, and about the role of fuzzy logic in our daily lives.

10.4 ILLUSTRATIVE EXAMPLES

The purpose of this closing section of the text is to illustrate applications of fuzzy set theory by a few simple examples of actual applications. A greater variety of examples can be found in [5].

As we mentioned earlier in this chapter, the greatest success of fuzzy set theory has been in the engineering area of control systems. *Fuzzy controllers* have been remarkably successful in dealing with control problems of a very broad spectrum, in terms of both performance and cost. Their high performance and low cost have often surpassed expectations.

One area in which fuzzy controllers quickly established an excellent reputation in the early 1990s is the broad area of *consumer products*. Among the earliest consumer products equipped with fuzzy controllers were *washing machines*. As our first example, we describe the basic ideas of the use of fuzzy controllers in this particular application. For pedagogical reasons, we substantially simplify the actual control problem. We consider only one controlled variable, the operating time of the machine for each given load of clothes; moreover, we omit all engineering details and implementation issues.

In a conventional washing machine, the time of each run is set by the user. In spite of this flexibility, it is difficult, even for fairly experienced users, to determine the right time. If insufficient time is set for a given load of clothes, they are not properly washed. If, on the other hand, the washing time is overly long, then time and energy are wasted and the machine as well as the clothes are unnecessarily worn out.

In overcoming these shortcomings of conventional washing machines, fuzzy control was found quite fitting. Different types of washing machines that employ fuzzy controllers, often referred to as *fuzzy washing machines*, are commercially available. Their control capabilities vary quite substantially. We now describe the operation of a very simple fuzzy controller, whose purpose is to determine the proper operating time of the washing machine for each load of clothes.

The operating time of a washing machine depends on two properties of each given load of clothes. First, it depends on how dirty the clothes are. Second, it depends on the type of soil. In a fuzzy washing machine, the *degree of dirtiness* is measured by a special sensor via the degree of water transparency. The less transparent the water, the dirtier the clothes. The type of soil is determined by measuring the time needed, after the machine has started, to reach a state in which the water transparency remains virtually constant. This time, which is called a saturation time, is different for different types of soil. For example, it is visibly shorter for muddy clothes than for oily clothes.

Assume that the degree of dirtiness, d, measured by the reduced transparency of water, is expressed by a number in the interval $[0, d_{max}]$, where d_{max} is some positive number that depends on the chosen measurement unit. Assume further that we want the fuzzy controller to deal only with three levels of dirtiness, expressed in natural language as *high, medium*, and *low*. Then it is reasonable to represent these linguistic terms by the trapezoidal-shape fuzzy numbers shown in Fig. 10.1. These numbers are states of a linguistic variable representing *dirtiness*, whose base variable is d (expressed by the reduction in water transparency). Let this linguistic variable be denoted by $\mathcal{D}$.

Assume now that the saturation time, s, estimated by the rate of increase in the degree of dirtiness, is expressed by a number in the interval $[0, s_{max}]$. Again, we want the fuzzy controller to deal only with three distinctions, expressed as *short, medium*, and *long* saturation times. Assume further that these linguistic terms are represented by the triangular fuzzy numbers shown in Fig. 10.2. These numbers are states of a linguistic variable representing *saturation time* whose base variable is s (saturation time expressed in terms of real numbers). Let this linguistic variable be denoted by $\mathcal{S}$.

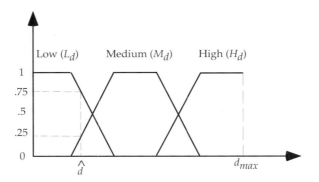

Figure 10.1   Fuzzy numbers representing three
levels of dirtiness: *low* (L_d), *medium* (M_d), *high* (H_d).
These are states of linguistic variable $\mathcal{D}$.

Intuitively, the required washing time should be some mathematical function of the degree of dirtiness and the saturation time. However, it is virtually impossible to determine this function exactly. By using a fuzzy controller we can approximate this function with relative ease on the basis of human intuition and experience. To do that we need to define another linguistic variable representing the required washing time. Let us denote it by $\mathcal{T}$ and its base variable by t.

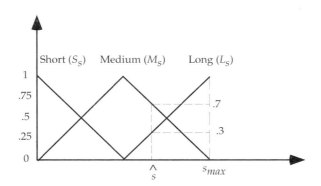

Figure 10.2   Fuzzy numbers representing *short*
(S_d), *medium* (M_s), *long* (L_d) saturation time. These
are states of linguistic variable $\mathcal{S}$.

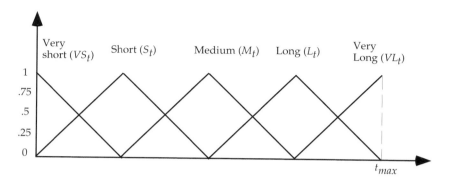

Figure 10.3   Fuzzy numbers characterizing the required washing time. These are states of linguistic variable $\mathcal{T}$.

Assume that $t \in [0, t_{max}]$ and that we want the fuzzy controller to deal with the five distinctions characterizing the required washing time expressed in natural language as *very short, short, medium, long,* and *very long*. A reasonable definition of fuzzy numbers representing these linguistic expressions is given in Fig. 10.3.

It is now relatively easy to express knowledge of experienced users of washing machines by conditional fuzzy propositions of the form

$$\text{If } \mathcal{D} = \square \text{ and } \mathcal{S} = \square, \text{ then } \mathcal{T} = \square$$

where appropriate states of the three linguistic variables are placed into the empty boxes for each particular proposition. Since variables $\mathcal{D}$ and $\mathcal{S}$ have three states each, the total number of possible ordered pairs of these states is nine. For each of these ordered pairs of states, we have to determine (using any available knowledge) an appropriate state of variable $\mathcal{T}$. This results in nine distinct conditional fuzzy propositions of the form shown above. These propositions are usually called *fuzzy inference rules* or *fuzzy if–then rules*. Examples of three of these rules are:

$$\text{If } \mathcal{D} = L_d \text{ and } \mathcal{S} = S_s, \text{ then } \mathcal{T} = VS_t$$
$$\text{If } \mathcal{D} = M_d \text{ and } \mathcal{S} = M_s, \text{ then } \mathcal{T} = M_t$$
$$\text{If } \mathcal{D} = H_d \text{ and } \mathcal{S} = L_s, \text{ then } \mathcal{T} = VL_t$$

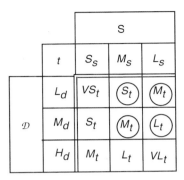

Figure 10.4   Inference rules for fuzzy washing machine.

In each rule, the states of $\mathcal{D}$ and $\mathcal{S}$ are called *antecedents* and the state of $\mathcal{T}$ is called the *consequent*.

    A convenient way of defining all required rules is the matrix shown in Fig. 10.4. Rows in the matrix correspond to states of variable $\mathcal{D}$, columns correspond to states of variable $\mathcal{S}$, and entries in the matrix are states of variable $\mathcal{T}$. Observe that the fuzzy inference rules defined by this matrix are in harmony with our common sense.

    The nine fuzzy inference rules represent knowledge upon which our fuzzy controller operates. The controller is connected to the washing machine as shown in Fig. 10.5. For given values of variables d and s, the controller determines the proper value of variable t (washing time) by executing the following steps.

    Step 1. When specific measured values of the input variables d and s, denoted by $\hat{d}$ and $\hat{s}$, are received by the controller at some predefined time, their compatibilities with the corresponding antecedents

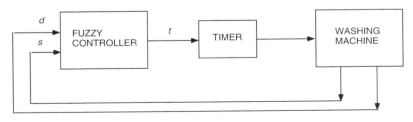

Figure 10.5   Fuzzy control of washing time.

of all inference rules are determined. For example, the measured value $\hat{d}$ shown Fig. 10.1 is compatible with $\mathcal{D} = L_d, M_d, H_d$ to the degrees of 0.75, 0.25, 0, respectively. Similarly, the measured value $\hat{s}$ shown in Fig. 10.2 is compatible with $\mathcal{S} = S_s, M_s, L_s$ to the degrees of 0, 0.7, 0.3, respectively. Only rules for which the compatibilities of the measured values with both antecedents are positive take place in determining the value of the controlled variable. These rules are usually referred to as *rules that fire*. In our example, the four rules which fire are identified in Fig. 10.4 by the circled entries. These rules are shown more explicitly in Fig. 10.6, together with the measured values $\hat{d}$ and $\hat{s}$ of our example.

Step 2. An inference is made by each rule that fires. To understand how this is done it is essential to realize that the rules attempt to approximate a function *f*, which is virtually impossible to be deter-

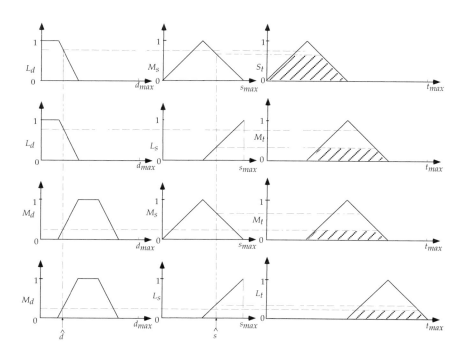

Figure 10.6   Inference rules that fire for $d = \hat{d}$ and $s = \hat{s}$.

mined exactly, by appropriate fuzzy numbers. In our example, the function has the form

$$t = f(d, s)$$

The approximation of a function by fuzzy numbers implicitly involves the extension principle. Given a particular fuzzy inference rule with two antecedents, their combination is a fuzzy set in $\mathbb{R}^2$ (a two-dimensional fuzzy set). If the antecedents are independent, as in our example, their combination is defined as the intersection of their cylindric extensions. The two-dimensional fuzzy set obtained in this way is then mapped into the consequent of the rule. Any antecedent whose degree of compatibility with a given measurement is less than 1 is truncated by this degree. The intersection of the cylindric extensions of these truncated antecedents is then truncated by the minimum degree of compatibility (assuming the standard fuzzy intersection) and this truncation is inherited in the consequent of the rule by the extension principle. This procedure is illustrated for our example in Fig. 10.6. Conclusions obtained by each of the individual inference rules for the measured values $\hat{d}$ and $\hat{s}$ are fuzzy sets whose membership functions are depicted by the shaded areas.

   Step 3. Given conclusions obtained by the individual fuzzy inference rules, we obtain the overall conclusion by taking the union of all the individual conclusions. In our example, illustrated in Fig. 10.6, the overall conclusion is the fuzzy set $C_{\hat{d},\hat{s}}$ whose membership function is defined for each $x \in [0, t_{max}]$ by the formula

$$C_{\hat{d},\hat{s}}(t) = \max\{\min[L_d(\hat{d}), M_s(\hat{s}), VS_t(t)], \min[L_d(\hat{d}), L_s(\hat{s}), S_t(t)]$$
$$\min[M_d(\hat{d}), M_s(\hat{s}), M_t(t)], \min[M_d(\hat{d}), L_s(\hat{s}), L_t(t)]\}$$

The graph of this function is shown in Fig. 10.7.

   Step 4. This last step in the operating procedure of fuzzy controllers is called *defuzzification*. Its purpose is to convert the fuzzy set representing the overall conclusion obtained in step 3 into a real number that, in some sense, best represents the fuzzy set. Although there are various defuzzification methods, each justified in some way, the most common method is to determine the value for which the area under the graph of the membership function is equally divided. This method is called a *center of gravity defuzzification method*. In general, given a fuzzy set A defined on the interval $[a_1, a_2]$, the center-of-gravity defuzzification, a, of A is defined by the formula

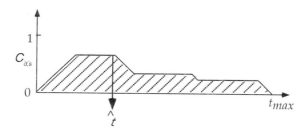

Figure 10.7 The fuzzy set which represents the overall conclusion for the measured values $\hat{d}$ and $\hat{s}$ and its defuzzified value $\hat{t}$.

$$a = \frac{\displaystyle\int_{a_1}^{a_2} xA(x)dx}{\displaystyle\int_{a_1}^{a_2} A(x)dx}$$

Applying this formula to our example, we obtain

$$\hat{t} = \frac{\displaystyle\int_{0}^{t_{max}} xC_{\hat{d},\hat{s}}(t)dt}{\displaystyle\int_{0}^{t_{max}} C_{\hat{d},\hat{s}}(t)dt}$$

This value, which is shown in Fig. 10.7, is the desirable operating time of the washing machine as determined by the fuzzy controller for conditions $\hat{d}$ and $\hat{s}$. The washing timer is set to this value.

Let us now describe an example of applying fuzzy set theory in a very different area, the area of *decision making*. This example deals with the problem of deciding which of several available houses to buy.

Assume that only five houses are available and we want to buy one. Assume further that we want to buy an attractive house, provided that its price, location (expressed in terms of the distance to work for the buyer), real estate taxes, and the quality of the associated school system are acceptable. All of these criteria can be expressed in terms of appropriate fuzzy sets.

Let the five houses be denoted by the labels a, b, c, d, e. Their prices, distances to work, and real estate taxes are specified in Table

10.1. Fuzzy sets *acceptable price* (*P*), *acceptable distance* (*D*), and *acceptable tax* (*T*), as defined by the buyer, are given in Fig. 10.8. Also shown in this figure are membership grades of the five houses in these sets. These membership grades are listed in Table 10.2. Membership grades of the five houses in the remaining two fuzzy sets—the set of *attractive houses* and the set of houses associated with *school systems with acceptable quality* are also listed in the table; these membership grades are determined by subjective judgment of the buyer.

TABLE 10.1   BASIC INFORMATION ABOUT FIVE HOUSES

House	Price (in thousand of dollars)	Distance to Work (in miles)	Taxes (in hundreds of dollars)
a	120	30	15
b	70	5	25
c	80	10	25
d	90	25	20
e	100	20	20

Membership grades of the houses in each of the five fuzzy sets express the degrees of acceptability of the houses according to the criterion represented by the fuzzy set. Since the buyer is interested in satisfying all the criteria, the degrees of overall acceptance of the individual houses are equivalent to their membership grades in the intersection of the five fuzzy sets. These membership grades, determined by the standard fuzzy intersection, are shown in the last column in Table 10.2. The house to buy is the one with the highest degree of overall acceptability, which is house *e*.

TABLE 10.2   DECISION-MAKING EXAMPLE: A SUMMARY

House	Degree of Acceptability of the Houses According to the Individual Criteria					Degree of Overall Acceptability
x	*P(x)*	*D(x)*	*T(x)*	*A(x)*	*S(x)*	
a	0.53	0.33	0.83	1	0.5	0.33
b	0.87	1	0.5	0.5	1	0.5
c	0.8	1	0.5	0.6	1	0.5
d	0.73	0.5	0.67	0.2	0.9	0.2
e	0.67	0.67	0.67	0.8	0.9	0.67

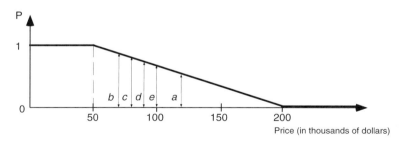

(a) Acceptable price

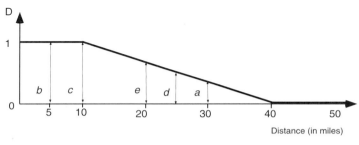

(b) Acceptable distance

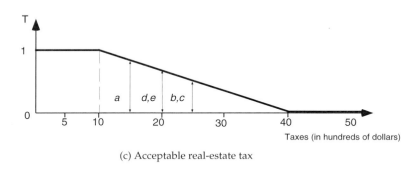

(c) Acceptable real-estate tax

Figure 10.8   Fuzzy sets defined by the buyer of a house.

Let us now describe a very simple example of the use of fuzzy set theory in fuzzy retrieval. The example involves two binary fuzzy relations. One of them describes degrees of relevance of given index terms (key words) to given documents; it is called a *relevance relation*. The other one, called a *fuzzy thesaurus*, describes for each pair of index terms the degree of closeness in meanings of the index terms involved.

For the sake of simplicity, we consider the set

$$D = \{d_1, d_2, ..., d_{15}\}$$

of only 15 documents, and the set

$$T = \{t_1, t_2, ..., t_7\}$$

of the following index terms:

t_1—fuzzy number
t_2 —interval arithmetic
t_3—fuzzy arithmetic
t_4—convexity
t_5—fuzzy relation
t_6—fuzzy set
t_7 —fuzzy proposition

We assume that the relevance of these index terms to the individual documents in set D has been determined and is expressed by the matrix

	d_1	d_2	d_3	d_4	d_5	d_6	d_7	d_8	d_9	d_{10}	d_{11}	d_{12}	d_{13}	d_{14}	d_{15}
t_1	0.2	0	0	0	0	0	0	0.7	0	0	0	0	0	0	1
t_2	0	0	0.3	0	0	0.9	1	1	0	0	0	0	0	0	0
t_3	0	0	0	0	1	0	0	0	0	0.6	0	0	0	0	0
t_4	0	0	0	0	0	0	0	0	0	0	0	0	0.2	0	1
t_5	1	0.7	0	0.5	0	0	0	0	0.5	0	0.9	0.5	0	0.5	0
t_6	0	0	0	0	1	1	1	0	0	0	0	0	0	0	0
t_7	0	1	1	0.5	0	0	0	0	0.4	0.8	0.4	0.2	1	1	0

with $\mathbf{R} =$ at the left.

representing a fuzzy relevance relation R on $T \times D$. We also assume that the closeness in meanings of the index terms has been determined and is expressed by the matrix

$$
\mathbf{M} = \begin{array}{c} \\ t_1 \\ t_2 \\ t_3 \\ t_4 \\ t_5 \\ t_6 \\ t_7 \end{array}
\begin{array}{ccccccc}
t_1 & t_2 & t_3 & t_4 & t_5 & t_6 & t_7 \\
\left[\begin{array}{ccccccc}
1 & 0.5 & 1 & 0.3 & 0 & 0.8 & 0.5 \\
0.5 & 1 & 0.7 & 0.1 & 0 & 0 & 0 \\
1 & .07 & 1 & 0 & 0 & 0.2 & 0 \\
0.3 & 0.1 & 0 & 1 & 0 & 0 & 0 \\
0 & 0 & 0 & 0 & 1 & 0.5 & 0 \\
0.8 & 0 & 0.2 & 0 & 0.5 & 1 & 0.9 \\
0.5 & 0 & 0 & 0 & 0 & 0.9 & 1
\end{array}\right]
\end{array}
$$

representing a fuzzy relation M on $T \times T$ (a fuzzy mini-thesaurus).

In fuzzy information retrieval, any *user's inquiry* is expressed by a fuzzy set I on T. The fuzzy set

$$
\begin{array}{ccccccc}
t_1 & t_2 & t_3 & t_4 & t_5 & t_6 & t_7 \\
\end{array}
$$
$$
\mathbf{I} = [\,0, 0.7, 1, 0.2, 0, 0, 0\,]
$$

expressed here by a vector of membership grades, is a possible user's inquiry in our example. When we compose I with the fuzzy thesaurus M, we obtain an augmented inquiry, A, which may extend the original inquiry by associated index terms. That is,

$$
A = I \circ M
$$

In our example,

$$
\begin{array}{ccccccc}
t_1 & t_2 & t_3 & t_4 & t_5 & t_6 & t_7 \\
\end{array}
$$
$$
\mathbf{A} = [\,1, 0.7, 1, 0.2, 0, 0.2, 0\,]
$$

The retrieved documents are then characterized by a fuzzy set Q on D, which is obtained by composing the augmented inquiry A with the relevance relation R. That is,

$$
Q = A \circ R
$$

In our example,

$$
\begin{array}{ccccccccccccccc}
d_1 & d_2 & d_3 & d_4 & d_5 & d_6 & d_7 & d_8 & d_9 & d_{10} & d_{11} & d_{12} & d_{13} & d_{14} & d_{15} \\
\end{array}
$$
$$\mathbf{Q} = [\ 0.2,\ \ 0,\ \ 0.3,\ \ 0,\ \ 1,\ \ 0.7, 0.7, 0.7,\ \ 0,\ \ 0.6,\ \ 0,\ \ \ 0,\ \ 0.2,\ \ 0,\ \ \ 1\ ]$$

This set characterizes the degree to which each document matches with the user's interest as expressed by his or her inquiry. It is now up to the user to decide whether to inspect all documents captured by the support of this set or to consider only documents captured by some α-cut of it.

References for Applications

[1] **Billot, A.** [**1992**]. *Economic Theory of Fuzzy Equilibria: An Axiomatic Analysis*. Springer-Verlag, New York.

[2] **Driankov, D., H. Hellendoorn, and M. Reinfrank** [**1993**]. *An Introduction to Fuzzy Control*. Springer-Verlag, New York.

[3] **Hall, L. O., and A. Kandel** [**1986**]. *Designing Fuzzy Expert Systems*. Verlag TÜV Rheinland, Cologne, Germany.

[4] **Kickert, W. J.** [**1978**]. *Fuzzy Theories on Decision-Making: Frontiers in Systems Research*. Martinus Nijhoff, Leiden, The Netherlands.

[5] **Klir, G. J., and B. Yuan** [**1995**]. *Fuzzy Sets and Fuzzy Logic: Theory and Applications*. Prentice Hall PTR, Upper Saddle River, NJ.

[6] **Klir, G. J., and B. Yuan** (ed.) [**1996**]. *Fuzzy Sets, Fuzzy Logic, and Fuzzy Systems: Selected Papers by Lotfi A. Zadeh*. World Scientific, Singapore.

[7] **Kosko, B.** [**1993**]. *Fuzzy Thinking*. Hyperion, New York.

[8] **Miyamoto S.** [**1990**]. *Fuzzy Sets in Information Retrieval and Cluster Analysis*. Kluwer, Boston, MA.

[9] **Pal, S. K., and D. K. Dutta Majumder** [**1986**]. *Fuzzy Mathematical Approach to Pattern Recognition*. John Wiley, New York.

[10] **Petry, F. E.** [**1996**]. *Fuzzy Databases: Principles and Applications*. Kluwer, Boston, MA.

[11] **Ross, T. J.** [**1995**]. *Fuzzy Logic with Engineering Applications*. McGraw-Hill, New York.

[12] **Rouvray, D. H.** (ed.) **[1997]**. *Fuzzy Logic in Chemistry*. Academic Press, San Diego, CA.

[13] **Yager, R. R., and D. P. Filev [1994]**. *Essentials of Fuzzy Modeling and Control*. John Wiley, New York.

[14] **Yager, R. R., S. Ovchinnikov, R. M. Tong, and H. T. Nguyen** (eds.) **[1987]**. *Fuzzy Sets and Applications—Selected Papers by L. A. Zadeh*. John Wiley, New York.

[15] **Zimmermann, H. J. [1987]**. *Fuzzy Sets, Decision Making, and Expert Systems*. Kluwer, Boston, MA.

[16] **Zimmermann, H. J. [1991]**. *Fuzzy Set Theory—and Its Applications*. (2nd rev. ed.). Kluwer, Boston, MA.

INDEX

THE
APOTHECARY
in Eighteenth-Century
WILLIAMSBURG

Being an Account of his medical and
chirurgical Services, as well as of
his trade Practices as a Chymist

Williamsburg Craft Series

WILLIAMSBURG
Publiſhed by *Colonial Williamſburg*

MMI